Flame Retardant and Thermally Insulating Polymers

Edited by Yanfei Xu

Published in London, United Kingdom

IntechOpen

Supporting open minds since 2005

Flame Retardant and Thermally Insulating Polymers
http://dx.doi.org/10.5772/intechopen.91119
Edited by Yanfei Xu

Contributors
Chanchal Kumar Kundu, Zhiwei Li, Lei Song, Yuan Hu, Sanjiv Sonakaria, Hyun-Joong Kim, Daniel Dobslaw, Bertram Kuch, Christine Woiski, Martina Kiel, Devaraju Subramnai, Muthukaruppan Alagar, Krishnamoorthy Krishnadevi, Yanfei Xu

First published in London, United Kingdom, 2021 by IntechOpen
IntechOpen is the global imprint of INTECHOPEN LIMITED, registered in England and Wales, registration number: 11086078, 5 Princes Gate Court, London, SW7 2QJ, United Kingdom
Printed in Croatia

British Library Cataloguing-in-Publication Data
A catalogue record for this book is available from the British Library

Additional hard and PDF copies can be obtained from orders@intechopen.com

Flame Retardant and Thermally Insulating Polymers
Edited by Yanfei Xu
p. cm.
Print ISBN 978-1-83968-714-3
Online ISBN 978-1-83968-715-0
eBook (PDF) ISBN 978-1-83968-716-7

For EU product safety concerns:
IN TECH d.o.o., Prolaz Marije Krucifikse Kozulić 3, 51000 Rijeka, Croatia, info@intechopen.com or visit our website at intechopen.com.

Meet the editor

Yanfei Xu joined the Mechanical and Industrial Engineering Department at the University of Massachusetts (UMass) Amherst as an assistant professor in 2019. She became an adjunct assistant professor in the Chemical Engineering Department in the fall of 2020. Before joining UMass, she was a postdoctoral researcher in the Mechanical Engineering Department at the Massachusetts Institute of Technology (MIT). Before joining MIT, she passionately pursued GENIUS research funded by the European Commission and secured Marie-Curie Fellowships with major European institutions and companies. Dr. Xu received a Ph.D. in Organic Chemistry from Nankai University, China.

Contents

Preface

Polymers have entered almost every aspect of modern life, thanks to their unparalleled properties such as lightweight, chemical resistance, electrical insulation, and easy processing, to name a few [1, 2]. However, common polymers are flammable because of their chemical structures, which consist of hydrogen and carbon atoms [3–5]. Inflammability properties in polymers are important for specific applications such as building insulation, aerospace parts, and firefighter uniforms [6]. To address polymer flammability issues, adding flame retardants into polymers is common [5, 7–13]. However, for traditional flame-retardant additives, high loading is usually needed to meet flame retardancy demands, which can lead to deteriorated mechanical properties and environmental issues [14–16]. Fire protection becomes crucial. Developing novel sustainable flame retardants is important [17, 18].

In addition to high flame retardancy, polymers with superior resistance to heat are essential for preventing burn injury under extreme conditions [19–23]. Common thermal insulation materials work by reducing conductive heat flow and, to a lesser extent, convective heat flow [24]. Radiant barriers and reflective insulation systems work by reducing radiant heat gain [24]. Thermally insulative materials could reduce the heat release rate [6]. Developing multifunctional polymers with excellent flame retardancy and superior thermal insulation is of fundamental interest and technological importance [10]. Thermal insulation properties can be improved by manipulating porous nanostructures in polymers to reduce solid conductivity, gaseous conductivity, and radiation heat transfer [25]. Advanced multifunctional polymers with low thermal conductivity on the order of 0.01 W m-1 K-1 have been developed [9, 19, 22, 26–34]. Various polymer foams, polymeric sponges, and composite aerogels with good thermal insulation and fireproof properties have been reported [10, 19, 22, 27, 35–44]. Formed char materials can also act as thermal insulation layers [45–47]. The thermal conductivity of char layers varies along with the evolution of the intumescent structures [47, 48]. Further investigation into the relationships between flame retardancy behaviors and polymer structures is needed [49]. Boosting flame retardancy and thermal insulation performance in polymers without deteriorating their mechanical property (e.g., compressive strength) and releasing toxic gases/products is highly desired [50–57].

Over the past decades, different flame retardants have been developed [6, 7, 20, 28, 33, 58–60]. Heat transfer influences processes in the ignition, growth, spread, decay, and extinction of fire [61, 62]. Flame retardants act either in the vapor phase or condensed phase to inhibit or to stop combustion processes through a chemical and/or physical mechanism [5, 60]. The flammability of polymer is also dependent on fire conditions — the transport of oxygen from the environment to the burning surface [26, 62, 63]. The polymer flammability behaviors have been explored by their ignitability, flame-spread rate, and heat release characterizations [64, 65]. Depending on the targeted polymer application, one or more specific flammability criteria need to be satisfied. Fame retardancy properties can be evaluated by various tests, which include limiting oxygen index, vertical burning test, cone calorimeter test, among others [66].

There are pioneering books and reviews for flame-retardant polymers. This book presents recent advances in the development of eco-friendly, flame-retardant, and thermally insulative polymer-based materials. It focuses not only on the developments of new eco-friendly, nontoxic, and high-performance flame retardants, but it also examines flame retardant behaviors in polymers. The introductory chapter "Flame Retardant and Thermally Insulating Polymers" highlights that, in addition to having superior flame retardancy, eco-friendly polymer-based materials with extraordinary thermal insulation and excellent mechanical strength are highly desired for unforeseen and existing applications such as building insulation and vehicle parts. Chapter 2, "Nanoscale Configuration of Clay-Interlayer Chemistry: A Precursor to Enhancing Flame Retardant Properties", presents the fundamentals of flame retardant behaviors and highlights how flame propagations are suppressed by chemical reactions between substances. Chapter 3, "Development of Halogen Free Sustainable Polybenzoxazine Matrices and Composites for Flame Retardant Applications", focuses on the design and synthesis of non-halogen, environmentally friendly, bio-based flame retardants. To evaluate these halogen-free flame retardant properties, the chapter presents limiting oxygen index and UL 94 vertical flammability test results. Chapter 4, "Plant Uptake, Translocation and Metabolism of PBDEs in Plants", emphasizes health and environmental issues caused by brominated flame retardants and highlights the development of innovative flame retardants. Chapter 5, "Flame Retardant Treatments of Nylon Textiles: A Shift towards Eco-Friendly Approaches," addresses how to design and develop sustainable, efficient, and durable flame retardants for nylon textiles, which are some of the most widely used polymers for industrial uses.

I would like to thank all the authors for their individual chapter contributions that made this book possible. I would also like to thank the reviewers who gave so much of their effort and time to review chapters, as well as the editors at IntechOpen who provided great support throughout the publication of this book.

Yanfei Xu
Mechanical and Industrial Engineering Department,
Chemical Engineering Department,
University of Massachusetts Amherst,
USA

References

[1] Peplow, M., *The plastics revolution: how chemists are pushing polymers to new limits.* Nature, 2016. **536**(7616): p. 266-268.

[2] Sperling, L.H., *Introduction to Physical Polymer Science, Fourth Edition.* John Wiley & Sons, Inc., 2005.

[3] Wehrmann, F., *Organic polymer chemistry, K. J. Saunders, Chapman and Hall, London, 1973, 473 pp.* Journal of Polymer Science: Polymer Letters Edition, 1975. **13**(10): p. 634-635.

[4] Camino, G. and L. Costa, *Performance and mechanisms of fire retardants in polymers—A review.* Polymer Degradation and Stability, 1988. **20**(3): p. 271-294.

[5] Camino, G., L. Costa, and M.P. Luda di Cortemiglia, *Overview of fire retardant mechanisms.* Polymer Degradation and Stability, 1991. **33**(2): p. 131-154.

[6] Laoutid, F., et al., *New prospects in flame retardant polymer materials: From fundamentals to nanocomposites.* Materials Science and Engineering: R: Reports, 2009. **63**(3): p. 100-125.

[7] Vahidi, G., et al., *Advancements in traditional and nanosized flame retardants for polymers—A review.* Journal of Applied Polymer Science, 2021. **138**(12): p. 50050.

[8] Zhu, Z.-M., et al., *Highly Flame Retardant Expanded Polystyrene Foams from Phosphorus–Nitrogen–Silicon Synergistic Adhesives.* Industrial & Engineering Chemistry Research, 2017. **56**(16): p. 4649-4658.

[9] Chen, H.-B., et al., *Highly Efficient Flame Retardant Polyurethane Foam with Alginate/Clay Aerogel Coating.* ACS Applied Materials & Interfaces, 2016. **8**(47): p. 32557-32564.

[10] Bourbigot, S. and S. Duquesne, *Fire retardant polymers: recent developments and opportunities.* Journal of Materials Chemistry, 2007. **17**(22): p. 2283-2300.

[11] Bourbigot, S., S. Duquesne, and C. Jama, *Polymer Nanocomposites: How to Reach Low Flammability?* Macromolecular Symposia, 2006. **233**(1): p. 180-190.

[12] Hornsby, P.R., *Fire retardant fillers for polymers.* International Materials Reviews, 2001. **46**(4): p. 199-210.

[13] M Le Bras, S.B., G Camino, R Delobel, *Fire Retardancy of Polymers*. 1998: Woodhead Publishing

[14] Wang, X., et al., *Carbon-family materials for flame retardant polymeric materials.* Progress in Polymer Science, 2017. **69**: p. 22-46.

[15] Morgan, A.B. and C.A. Wilkie, *The Non-Halogenated Flame Retardant Handbook*. 2014, Somerset, UNITED STATES: John Wiley & Sons, Incorporated.

[16] Lu, S.-Y. and I. Hamerton, *Recent developments in the chemistry of halogen-free flame retardant polymers.* Progress in Polymer Science, 2002. **27**(8): p. 1661-1712.

[17] Zhang, Y., et al., *Eco-friendly flame retardant and electromagnetic interference shielding cotton fabrics with multi-layered coatings.* Chemical Engineering Journal, 2019. **372**: p. 1077-1090.

[18] Zhao, H.-B., et al., *A flame-retardant-free and thermo-cross-linkable copolyester: Flame-retardant and anti-dripping mode of action.* Polymer, 2014. **55**(10): p. 2394-2403.

[19] Shi, H.-G., et al., *Multifunctional Flame-Retardant Melamine-Based Hybrid Foam for Infrared Stealth,*

Thermal Insulation, and Electromagnetic Interference Shielding. ACS Applied Materials & Interfaces, 2021.

[20] Lee, H., et al., *Super-insulating, flame-retardant, and flexible poly(dimethylsiloxane) composites based on silica aerogel.* Composites Part A: Applied Science and Manufacturing, 2019. **123**: p. 108-113.

[21] Tran, P., Q.T. Nguyen, and K.T. Lau, *Fire performance of polymer-based composites for maritime infrastructure.* Composites Part B: Engineering, 2018. **155**: p. 31-48.

[22] Wicklein, B., et al., *Thermally insulating and fire-retardant lightweight anisotropic foams based on nanocellulose and graphene oxide.* Nature Nanotechnology, 2015. **10**(3): p. 277-283.

[23] Stahl, T., et al., *Thermo-hygric properties of a newly developed aerogel based insulation rendering for both exterior and interior applications.* Energy and Buildings, 2012. **44**: p. 114-117.

[24] Office of Energy Saver, O.o.E.E.R.E., *Radiant Barriers*. https://www.energy.gov/energysaver/weatherize/insulation/radiant-barriers.

[25] Lu, X., et al., *Thermal Conductivity of Monolithic Organic Aerogels.* Science, 1992. **255**(5047): p. 971.

[26] Gibson, A.P.M.a.A.G., *Fire Properties of Polymer Composite Materials*. 2006: Springer.

[27] Ma, Z., et al., *Bioinspired, Highly Adhesive, Nanostructured Polymeric Coatings for Superhydrophobic Fire-Extinguishing Thermal Insulation Foam.* ACS Nano, 2021.

[28] Akdogan, E., et al., *Synergistic effects of expandable graphite and ammonium pentaborate octahydrate on the flame-retardant, thermal insulation, and mechanical properties of rigid polyurethane foam.* Polymer Composites, 2020. **41**(5): p. 1749-1762.

[29] Xiao, Y., et al., *Compressible, Flame-Resistant and Thermally Insulating Fiber-Reinforced Polybenzoxazine Aerogel Composites.* Materials (Basel, Switzerland), 2020. **13**(12): p. 2809.

[30] Cao, Z.-J., et al., *Polyurethane foams with functionalized graphene towards high fire-resistance, low smoke release, superior thermal insulation.* Chemical Engineering Journal, 2019. **361**: p. 1245-1254.

[31] Wang, S.-X., et al., *Inherently flame-retardant rigid polyurethane foams with excellent thermal insulation and mechanical properties.* Polymer, 2018. **153**: p. 616-625.

[32] Wei, H., et al., *Benzotriazole-based conjugated microporous polymers as efficient flame retardants with better thermal insulation properties.* Journal of Materials Chemistry A, 2018. **6**(18): p. 8633-8642.

[33] Han, Y., et al., *Flame Retardant, Heat Insulating Cellulose Aerogels from Waste Cotton Fabrics by in Situ Formation of Magnesium Hydroxide Nanoparticles in Cellulose Gel Nanostructures.* ACS Sustainable Chemistry & Engineering, 2015. **3**(8): p. 1853-1859.

[34] Jelle, B.P., *Traditional, state-of-the-art and future thermal building insulation materials and solutions – Properties, requirements and possibilities.* Energy and Buildings, 2011. **43**(10): p. 2549-2563.

[35] Luo, M., et al., *Enhanced Thermal Insulation and Flame-Retardant Properties of Polyvinyl Alcohol-Based Aerogels Composited with Ammonium Polyphosphate and Chitosan.* International Journal of Polymer Science, 2021. **2021**: p. 5555916.

[36] Zhu, J., et al., *Thermal insulation and flame retardancy of attapulgite reinforced gelatin-based composite aerogel*

with enhanced strength properties. Composites Part A: Applied Science and Manufacturing, 2020. **138**: p. 106040.

[37] Wang, T., et al., *An ultralow-temperature superelastic polymer aerogel with high strength as a great thermal insulator under extreme conditions.* Journal of Materials Chemistry A, 2020. **8**(36): p. 18698-18706.

[38] Jin, F.-L., et al., *Recent Trends of Foaming in Polymer Processing: A Review.* Polymers, 2019. **11**(6): p. 953.

[39] Rizvi, A., R.K.M. Chu, and C.B. Park, *Scalable Fabrication of Thermally Insulating Mechanically Resilient Hierarchically Porous Polymer Foams.* ACS Applied Materials & Interfaces, 2018. **10**(44): p. 38410-38417.

[40] Yu, Z.-L., et al., *Fire-Retardant and Thermally Insulating Phenolic-Silica Aerogels.* Angewandte Chemie International Edition, 2018. **57**(17): p. 4538-4542.

[41] Xie, H., et al., *Study on flame retarded flexible polyurethane foam/alumina aerogel composites with improved fire safety.* Chemical Engineering Journal, 2017. **311**: p. 310-317.

[42] Zhao, H.-B., M. Chen, and H.-B. Chen, *Thermally Insulating and Flame-Retardant Polyaniline/Pectin Aerogels.* ACS Sustainable Chemistry & Engineering, 2017. **5**(8): p. 7012-7019.

[43] Stapleton, H.M., et al., *Detection of Organophosphate Flame Retardants in Furniture Foam and U.S. House Dust.* Environmental Science & Technology, 2009. **43**(19): p. 7490-7495.

[44] Wang, J.Q. and W.K. Chow, *A brief review on fire retardants for polymeric foams.* Journal of Applied Polymer Science, 2005. **97**(1): p. 366-376.

[45] *Combustion of Polymers, Retardancy due to char formation.*

[46] Anderson, C.E., D.E. Ketchum, and W.P. Mountain, *Thermal Conductivity of Intumescent Chars.* Journal of Fire Sciences, 1988. **6**(6): p. 390-410.

[47] S. Bourbigot, S.D.a.J.-M.L., *Modeling of Heat Transferof a Polypropylene-BasedIntumescent System duringCombustion.* J. Fire Sci., 1999. **17**: p. 42.

[48] Camino, G., L. Costa, and G. Martinasso, *Intumescent fire-retardant systems.* Polymer Degradation and Stability, 1989. **23**(4): p. 359-376.

[49] Weil, E.D. and S.V. Levchik, *Flame Retardants for Plastics and Textiles : Practical Applications*. 2015, München, GERMANY: Hanser.

[50] Kim, Y., S. Lee, and H. Yoon, *Fire-Safe Polymer Composites: Flame-Retardant Effect of Nanofillers.* Polymers, 2021. **13**(4).

[51] Jin, H., et al., *Ultralight and Hydrophobic Palygorskite-based Aerogels with Prominent Thermal Insulation and Flame Retardancy.* ACS Applied Materials & Interfaces, 2020. **12**(10): p. 11815-11824.

[52] Won, J.-P., et al., *Eco-friendly fireproof high-strength polymer cementitious composites.* Construction and Building Materials, 2012. **30**: p. 406-412.

[53] Gu, P. and R.J. Asaro, *Structural buckling of polymer matrix composites due to reduced stiffness from fire damage.* Composite Structures, 2005. **69**(1): p. 65-75.

[54] Gibson, A.G., et al., *Modelling residual mechanical properties of polymer composites after fire.* Plastics, Rubber and Composites, 2003. **32**(2): p. 81-90.

[55] Covaci, A., S. Voorspoels, and J. de Boer, *Determination of brominated flame retardants, with emphasis on*

polybrominated diphenyl ethers (PBDEs) in environmental and human samples—a review. Environment International, 2003. **29**(6): p. 735-756.

[56] Rahman, F., et al., *Polybrominated diphenyl ether (PBDE) flame retardants.* Science of The Total Environment, 2001. **275**(1): p. 1-17.

[57] Mouritz, A.P. and Z. Mathys, *Mechanical properties of fire-damaged glass-reinforced phenolic composites.* Fire and Materials, 2000. **24**(2): p. 67-75.

[58] Morgan, A.B. and J.W. Gilman, *An overview of flame retardancy of polymeric materials: application, technology, and future directions.* Fire and Materials, 2013. **37**(4): p. 259-279.

[59] Thirumal, M., et al., *Effect of a Nanoclay on the Mechanical, Thermal and Flame Retardant Properties of Rigid Polyurethane Foam.* Journal of Macromolecular Science, Part A, 2009. **46**(7): p. 704-712.

[60] Price, A.R.H.D., *Fire Retardant Materials*. Woodhead Publishing.

[61] NIST, *Fire dynamics.* https://www.nist.gov/el/fire-research-division-73300/firegov-fire-service/fire-dynamics.

[62] *Introduction Polymers and Fire*, in *Fire Retardancy of Polymers: New Strategies and Mechanisms*, T.R. Hull and B.K. Kandola, Editors. 2009, The Royal Society of Chemistry. p. 1-14.

[63] Camino, G., et al., *The oxygen index method in fire retardance studies of polymeric materials.* Journal of Applied Polymer Science, 1988. **35**(7): p. 1863-1876.

[64] Babu, K., et al., *A Review on the Flammability Properties of Carbon-Based Polymeric Composites: State-of-the-Art and Future Trends.* Polymers, 2020. **12**(7).

[65] Hirschler, M.M., *Flame retardants and heat release: review of traditional studies on products and on groups of polymers.* Fire and Materials, 2015. **39**(3): p. 207-231.

[66] Dasari, A., et al., *Recent developments in the fire retardancy of polymeric materials.* Progress in Polymer Science, 2013. **38**(9): p. 1357-1387.

Chapter 1

Introductory Chapter: Flame Retardant and Thermally Insulating Polymers

Yanfei Xu

1. Introduction

Flame retardant and thermally insulative polymers are of technological importance and fundamental interest [1, 2]. Polymers continue to infiltrate modern technologies such as aviation, automotive industry, building construction, electronics, to name a few, thanks to their unique combination of properties not available from any other known materials [3]. Polymers are lightweight, durable, easy to process, electrically insulative and corrosion resistant [4]. Common polymers are also thermally insulative [5]. However, polymers are combustible because of their chemical structures that are made up by carbon and hydrogen atoms [6–8]. To meet flammability standards, flame retardants for fireproof polymers have been developed [9, 10].

To protect human life and property, flame retardant polymers are generally made by adding flame retardants into polymers [11, 12]. There are drawbacks in common halogenated flame retardants, which associate with the release of toxic or corrosive by-products [13–18]. There are environment and health concerns caused by these released toxic gases [19]. Therefore, developing high-performance, non-toxic, low-cost, and environmentally friendly flame retardants are needed [19, 20]. Understanding mechanisms for fire retardancy is essential for developing new effective flame retardants. Improving fire retardant behaviors of polymers play key roles in future industrial applications such as furnishings, transportation products and building construction materials [9, 21–25].

Over the past decades, different flame retardants for polymers have been developed [20]. Mechanisms of polymer flame retardancy have been further investigated [9, 20, 26, 27]. Flame retardants have been generally broken into categories based on chemical compositions, which are grouped based on whether they contain bromine, chlorine, phosphorus, nitrogen, boron, or inorganic fillers (metals, etc.) [28]. It is widely recognized that all categories of flame retardants act either in vapor phase or condensed phase to inhibit or to stop combustion processes through a chemical and/or physical mechanism [29]. Flame retardants can interfere with combustion during a particular stage, e.g., during heating, pyrolysis, ignition, or flame spread [9, 20, 27, 30]. Flame retardants can either act chemically (reaction in the condensed or gas phase) and/or physically (by cooling, formation of a protective layer or fuel dilution) [17]. The polymer flammability properties have been investigated by their ignitability, flame spread and heat release characterizations [20]. Depending on the targeted application of polymers, one or more of specific flammability criteria (e.g. ASTM's fire and flammability standards) need to be satisfied [31].

In addition to flame retardancy, thermal insulation property of polymers can be another significant function that can defend targets against heat damage and save lives [32–35]. Fully understanding the flame retardance and thermal insulation

IntechOpen

mechanisms in polymers remain challenging. Advanced polymers with combined properties of flame retardancy, mechanical strength and heat insulation are needed. Such polymers will provide broader prospect in civil applications than single-function polymers, for example, building insulation applications [36–38].

This introductory chapter not only aims to present the current landscape flame retardant and thermally insulative polymers, but also highlights next generation of flame retardant and thermally insulative polymers for fire protections applications. This introductory chapter summarize fundamental interests and technological importance of flame retardant and thermally insulative polymers, which include principles of polymer flammability, theory of flame retardance, thermally insulative and fire-retardant polymers, and critical discussion and outlook.

2. History and perspective of flame retardant and thermally insulative polymers

2.1 Principles of polymer flammability

Three ingredients — heat, oxygen, and fuel — are required to initiate and continue a fire [12]. Polymer starts to degrade, when it is heated by external ignition sources and reaches a characteristic temperature [39]. The surrounding oxygen amount plays key roles on polymer surface decompositions (e.g., thermo-oxidative degradation of polymers and/or thermal degradation of polymers). The amount of oxygen for polymer decomposition depends on the specific polymer used. Combustible gases as fuels may be produced at a rate dependent upon polymer decomposition rate and diffuse to the flame front [12]. After ignition and removal of the ignition source, combustion could be self-propagating if there is sufficient heat generated and polymer can absorb enough heat to sustain its decomposition processes. Polymer combustion processes could involve vapor phase and condensed phase reactions [39].

2.2 Theory of flame retardance

2.2.1 Vapor phase flame inhibition

The combustion process of premixed methane-oxygen flame is well investigated [40, 41]. The methane oxygen system can be used as a model for studying more complex polymer flames [41, 42]. Methane combustion is a free-radical chain reaction, which mainly consists of propagation, chain branching, and termination processes [43]. Any flame-retardant material which either decreases the concentration of these chain carrying radicals or increases the rate of termination will inhibit the flame reaction. This is thought to be mechanisms by which vapor phase flame inhibitors [39, 44].

2.2.2 Condensed phase flame inhibition

Cooling and char barrier formation are two main modes in solid phase flame inhibition [39, 44, 45]. We discuss cooling mode first. One important cooling mode used in condensed phase flame inhibition is the use of materials which decompose endothermically in the pyrolysis zone of the burning polymer [39]. For example, due to polyvinyl alcohol's ability to endothermically form water molecules, polyvinyl alcohol is less flammable than the isomeric polyethylene oxide [39, 46]. During polymer burning processes, cooling mode can be also achieved by adding thermally conductive fillers into polymers [39]. Fillers have higher thermal conductivities than that of polymers. Fillers conduct heat better than polymers. Fillers conduct heat away from hot regions more efficiently than unfilled polymers. Fillers enable polymer-filler composites difficult to burn. However, fillers at high volume fraction loadings are needed for good cooling effect, which might lead to limited use [39, 47]. Dripping is another cooling mechanism [39, 48]. Polymers that drip easily during burning processes are more difficult to burn. For example, a regular candle will not burn with no wick, due to its high dripping tendency. This is because heat is dissipated from flaming areas. However, dripping could be a hazard by resulting in the spreading of a fire and thus of limited use [39, 49].

2.3 Thermally insulative polymers with high flame retardancy

Thermally insulative polymer-based materials with high flame retardancy are attracting significant attention [50]. This is because thermally insulative materials can protect overheating damage from burn injuries and save lives [34, 51–58]. There are drawbacks for available fire-resistant polymers. Some flame retardant polymers could be expensive [57]. Some flame retardant polymers have relatively low decomposition temperatures and decompose nearby 400°C [57]. Thus, highly thermally insulative, thermally stable and flame retardant polymer-based materials are desired for advanced thermal management applications [17, 50, 57].

Flame retardant and thermally insulative polymer-based composites have been developed [50, 56, 59–62]. For example, PC–PDMS copolymers have flame-retardant behaviors. Chars can prevent more volatile fuel production and serve as a thermal insulator preventing the temperature from rising [59]. When a specimen of PC–PDMS was in combustion, a lot of fine bubbles and char were formed. These fine bubbles are good for thermal insulation [59]. Moreover, silica particles in situ produced by thermal decomposition of PDMS mostly stay in char layers, which improve the quantity of oxidation-resistant char coatings [59]. The resulting bubble structures and silica materials in the char layer prevented volatile and flammable fuel production, which served as an effective thermal insulator [59]. Although there are progresses on developing flame retardant and thermally insulative polymer-based materials, further understanding flame retardancy and thermal insulation mechanisms will play key roles increating next generation of thermally insulative and flame retardant polymers with outstanding performance. With unique combined properties including simple manufacturing process, low cost, excellent thermal insulation, flame retardancy, superior physical and mechanical properties, thermally insulative and flame retardant polymers will provide new opportunities for existing and unforeseen applications.

2.4 Critical discussion and outlook

To achieve the high-performance fire retardancy of polymers, different strategies have been developed [63]. Followings are selected strategies.

1. By modifying the reaction scheme of pyrolysis of polymers to produce non-combustible, and/or non-volatile products that dilute the supply of oxygen [40, 63].

2. By stopping the combustion through dilution of the combustible gases, or the formation of a char which suppress the oxygen supply [63, 64].

3. By introducing active radical-trapping effects both in the gaseous phase and/or in the condensed phase [63, 65].

4. By reducing the thermal conductivity of the material to limit heat transfer [34, 63].

The different types of flame retardants in polymers based on halogens, heavy phosphorus-organic compounds and/or transition metals have shown good flame retardance performance [26]. However, toxic gases and smoke are formed during burning processes [66]. Environmental safety of flame retardants in polymers is a major issue [22, 63]. Fire-retardant polymer-based materials are desired to have high resistance to ignition, low combustion rate, retention of low flammability, acceptability in properties and appearance, no health safety and environmental issues, and little (or no) economic penalty.

Author details

Yanfei Xu
Mechanical and Industrial Engineering Department, Chemical Engineering Department, University of Massachusetts Amherst, Amherst, MA, USA

*Address all correspondence to: yanfeixu@umass.edu

References

[1] Yuan Hu, X.W., *Flame-retardant polymeric materials*. 2020: CRC Press.

[2] Bourbigot, S. and S. Duquesne, *Fire retardant polymers: recent developments and opportunities.* Journal of Materials Chemistry, 2007. **17**(22): p. 2283-2300.

[3] Peplow, M., *The plastics revolution: how chemists are pushing polymers to new limits.* Nature, 2016. **536**(7616): p. 266-8.

[4] Sperling, L.H., *Introduction to Physical Polymer Science, Fourth Edition.* John Wiley & Sons, Inc., 2005.

[5] Xu, Y., et al., *Nanostructured polymer films with metal-like thermal conductivity.* Nature Communications, 2019. **10**(1): p. 1771.

[6] Wehrmann, F., *Organic polymer chemistry, K. J. Saunders, Chapman and Hall, London, 1973, 473 pp.* Journal of Polymer Science: Polymer Letters Edition, 1975. **13**(10): p. 634-635.

[7] Hilado, C.J., *Flammability handbook for plastics*. 1998: CRC Press.

[8] Charles A. Wilkie, A.B.M., *Fire Retardancy of Polymeric Materials.* CRC Press, 2009.

[9] Vahidi, G., et al., *Advancements in traditional and nanosized flame retardants for polymers—A review.* Journal of Applied Polymer Science, 2021. **138**(12): p. 50050.

[10] M Le Bras, S.B., G Camino, R Delobel, *Fire Retardancy of Polymers*. 1998: Woodhead Publishing

[11] Kashiwagi, T., et al., *Nanoparticle networks reduce the flammability of polymer nanocomposites.* Nature Materials, 2005. **4**(12): p. 928-933.

[12] Khanna, Y.P. and E.M. Pearce, *Flammability of Polymers*, in *Applied Polymer Science*. 1985, American Chemical Society. p. 305-319.

[13] Birnbaum Linda, S. and F. Staskal Daniele, *Brominated flame retardants: cause for concern?* Environmental Health Perspectives, 2004. **112**(1): p. 9-17.

[14] Ames, A.B.a.B.N., *Flame-retardant additives as possible cancer hazards.* Science, 1977. **195**: p. 17-23.

[15] Osimitz, T.G., S. Kacew, and A.W. Hayes, *Assess flame retardants with care.* Science, 2019. **365**(6457): p. 992.

[16] Shaw, S.D., et al., *Halogenated flame retardants: do the fire safety benefits justify the risks?* Rev Environ Health, 2010. **25**(4): p. 261-305.

[17] Laoutid, F., et al., *New prospects in flame retardant polymer materials: From fundamentals to nanocomposites.* Materials Science and Engineering: R: Reports, 2009. **63**(3): p. 100-125.

[18] Blum, A., *The fire retardant dilemma.* Science, 2007. **318**(5848): p. 194-5.

[19] Kiliaris, C.D.P.a.P., *Polymer Green Flame Retardants*. Elsevier B.V. . 2014.

[20] Lu, S.-Y. and I. Hamerton, *Recent developments in the chemistry of halogen-free flame retardant polymers.* Progress in Polymer Science, 2002. **27**(8): p. 1661-1712.

[21] Lazar, S.T., T.J. Kolibaba, and J.C. Grunlan, *Flame-retardant surface treatments.* Nature Reviews Materials, 2020. **5**(4): p. 259-275.

[22] Xiong, P., et al., *A Review of Environmental Occurrence, Fate, and Toxicity of Novel Brominated Flame Retardants.* Environmental Science & Technology, 2019. **53**(23): p. 13551-13569.

[23] Price, A.R.H.a.D., *Advances in Fire Retardant Materials*. Woodhead Publishing. 2008.

[24] Wang, D.-Y., *Novel Fire Retardant Polymers and Composite Materials*. 2016, Cambridge, UNITED KINGDOM: Elsevier Science & Technology.

[25] Morgan, A.B. and C.A. Wilkie, *The Non-Halogenated Flame Retardant Handbook*. 2014, Somerset, UNITED STATES: John Wiley & Sons, Incorporated.

[26] Camino, G. and L. Costa, *Performance and mechanisms of fire retardants in polymers—A review.* Polymer Degradation and Stability, 1988. **20**(3): p. 271-294.

[27] Camino, G., L. Costa, and M.P. Luda di Cortemiglia, *Overview of fire retardant mechanisms.* Polymer Degradation and Stability, 1991. **33**(2): p. 131-154.

[28] Visakh, P.M., Arao, Yoshihiko, *Flame Retardants.* Springer, 2015.

[29] Kashiwagi, T., *Polymer combustion and flammability—Role of the condensed phase.* Symposium (International) on Combustion, 1994. **25**(1): p. 1423-1437.

[30] Kausar, A., et al., *Recent Developments in Different Types of Flame Retardants and Effect on Fire Retardancy of Epoxy Composite.* Polymer-Plastics Technology and Engineering, 2016. **55**(14): p. 1512-1535.

[31] Babrauskas, V. and R.D. Peacock, *c.* Fire Safety Journal, 1992. **18**(3): p. 255-272.

[32] Shi, H.-G., et al., *Multifunctional Flame-Retardant Melamine-Based Hybrid Foam for Infrared Stealth, Thermal Insulation, and Electromagnetic Interference Shielding.* ACS Applied Materials & Interfaces, 2021.

[33] Wicklein, B., et al., *Thermally insulating and fire-retardant lightweight anisotropic foams based on nanocellulose and graphene oxide.* Nature Nanotechnology, 2015. **10**(3): p. 277-283.

[34] Zhao, H.-B., M. Chen, and H.-B. Chen, *Thermally Insulating and Flame-Retardant Polyaniline/Pectin Aerogels.* ACS Sustainable Chemistry & Engineering, 2017. **5**(8): p. 7012-7019.

[35] Weil, E.D., *Flame Retardants for Plastics and Textiles 2E: Practical Applications*. 2015: Hanser Publications.

[36] Jelle, B.P., *Traditional, state-of-the-art and future thermal building insulation materials and solutions – Properties, requirements and possibilities.* Energy and Buildings, 2011. **43**(10): p. 2549-2563.

[37] Weil, E.D. and S.V. Levchik, *Flame Retardants for Plastics and Textiles : Practical Applications*. 2015, München, GERMANY: Hanser.

[38] *International Energy Agency. Technology Roadmap: Energy Efficient Building Envelopes*. 2013.

[39] Factor, A., *The chemistry of polymer burning and flame retardance.* Journal of Chemical Education, 1974. **51**(7): p. 453.

[40] Wall, L.A., *The Mechanisms of Pyrolysis, Oxidation, and Burning of Organic Materials*. Forgotten Books

[41] Westenberg, A.A. and R.M. Fristrom, *METHANE-OXYGEN FLAME STRUCTURE. IV. CHEMICAL KINETIC CONSIDERATIONS.* The Journal of Physical Chemistry, 1961. **65**(4): p. 591-601.

[42] Westenberg, R.M.F.a.A.A., *Flame structure*. McGraw-Hill.

[43] Korobeinichev, O.P., et al., *Chain-branching reactions in the processes of promotion and inhibition of hydrogen combustion.* Combustion, Explosion,

and Shock Waves, 2010. **46**(2): p. 140-148.

[44] Morgan, C.A.W.a.A.B., *Fire Retardancy of Polymeric Materials, Second Edition*. CRC Press.

[45] Wilkie, C.A., *Fire retardancy of polymers.* Materials Today, 2009. **12**(4): p. 46.

[46] Holland, B.J. and J.N. Hay, *The thermal degradation of poly(vinyl alcohol).* Polymer, 2001. **42**(16): p. 6775-6783.

[47] Rothon, R. and P. Hornsby, *Chapter 9 - Fire Retardant Fillers for Polymers*, in *Polymer Green Flame Retardants*, C.D. Papaspyrides and P. Kiliaris, Editors. 2014, Elsevier: Amsterdam. p. 289-321.

[48] Matzen, M., et al., *Influence of Flame Retardants on the Melt Dripping Behaviour of Thermoplastic Polymers.* Materials (Basel, Switzerland), 2015. **8**(9): p. 5621-5646.

[49] Liu, B.-W., et al., *Eco-friendly synergistic cross-linking flame-retardant strategy with smoke and melt-dripping suppression for condensation polymers.* Composites Part B: Engineering, 2021. **211**: p. 108664.

[50] Morgan, A.B. and J.W. Gilman, *An overview of flame retardancy of polymeric materials: application, technology, and future directions.* Fire and Materials, 2013. **37**(4): p. 259-279.

[51] Wi, S., et al., *Evaluation of environmental impact on the formaldehyde emission and flame-retardant performance of thermal insulation materials.* Journal of Hazardous Materials, 2021. **402**: p. 123463.

[52] Akdogan, E., et al., *Rigid polyurethane foams with halogen-free flame retardants: Thermal insulation, mechanical, and flame retardant properties.* Journal of Applied Polymer Science, 2020. **137**(1): p. 47611.

[53] Fan, B., et al., *Fabrication of Cellulose Nanofiber/AlOOH Aerogel for Flame Retardant and Thermal Insulation.* Materials, 2017. **10**(3).

[54] Wang, D., et al., *Biomimetic structural cellulose nanofiber aerogels with exceptional mechanical, flame-retardant and thermal-insulating properties.* Chemical Engineering Journal, 2020. **389**: p. 124449.

[55] Wei, H., et al., *Benzotriazole-based conjugated microporous polymers as efficient flame retardants with better thermal insulation properties.* Journal of Materials Chemistry A, 2018. **6**(18): p. 8633-8642.

[56] Kashiwagi, T., et al., *Flame retardant mechanism of silica gel/silica.* Fire and Materials, 2000. **24**(6): p. 277-289.

[57] Illeperuma, W.R.K., et al., *Fire-Resistant Hydrogel-Fabric Laminates: A Simple Concept That May Save Lives.* ACS Applied Materials & Interfaces, 2016. **8**(3): p. 2071-2077.

[58] Morgan, A.B., *The Future of Flame Retardant Polymers – Unmet Needs and Likely New Approaches.* Polymer Reviews, 2019. **59**(1): p. 25-54.

[59] Nodera, A. and T. Kanai, *Flame retardancy of a polycarbonate–polydimethylsiloxane block copolymer: The effect of the dimethylsiloxane block size.* Journal of Applied Polymer Science, 2006. **100**(1): p. 565-575.

[60] Zhu, Z., et al., *Ionic liquid and magnesium hydrate incorporated conjugated microporous polymers nanotubes with superior flame retardancy and thermal insulation.* Polymer, 2020. **194**: p. 122387.

[61] Wang, F., et al., *Monolithic nanoporous polymers bearing POSS*

moiety as efficient flame retardant and thermal insulation materials. Reactive and Functional Polymers, 2019. **143**: p. 104345.

[62] Wang, S.-X., et al., *Inherently flame-retardant rigid polyurethane foams with excellent thermal insulation and mechanical properties.* Polymer, 2018. **153**: p. 616-625.

[63] Zaikov, G.E. and S.M. Lomakin, *New Types of Ecologically Safe Flame-Retardant Polymer Systems*, in *Fire and Polymers II*. 1995, American Chemical Society. p. 186-198.

[64] Salmeia, K.A., et al., *An Overview of Mode of Action and Analytical Methods for Evaluation of Gas Phase Activities of Flame Retardants.* Polymers, 2015. 7(3).

[65] Guo, Z., Z. Wang, and Z. Fang, *Fabrication of 9,10-dihydro-9-oxa-10-phosphaphenanthrene-10-oxide-decorated fullerene to improve the anti-oxidative and flame-retardant properties of polypropylene.* Composites Part B: Engineering, 2020. **183**: p. 107672.

[66] Pethsangave, D.A., et al., *Novel Approach toward the Synthesis of a Phosphorus-Functionalized Polymer-Based Graphene Composite as an Efficient Flame Retardant.* ACS Sustainable Chemistry & Engineering, 2019. 7(13): p. 11745-11753.

Chapter 2

Nanoscale Configuration of Clay-Interlayer Chemistry: A Precursor to Enhancing Flame Retardant Properties

Sanjiv Sonkaria and Hyun Joong Kim

Abstract

Nanomaterials are proving to be pivotal to the evolution of controllable, cost-effective and environmentally safe technologies. An important concern is the impact of low-dimensional compositional materials and their ability to significantly reduce the hazardous nature of flame retardants that are reputably harmful through unchecked inhalation. While eco-friendly and recyclable alternatives are necessary requirements to function as replacements for the 'Next Generation' of flame retardants, the underlying 'Chemistry' at the nanoscale is unfolding unlocking vital clues enabling the development of more effective retardants. In this direction, the dimensional order of particles in naturally occurring nanoclay materials and their associated properties as composites are gaining increasing attention as important constituents of flame retardants. In this review, we examine closer the compositional importance of intercalated/exfoliated nanoclay networks essential to retardant functionality exploring the chemical significance and discussing underlying mechanisms where possible.

Keywords: nanoclay, flame retardant, nanocomposite, polymer–layered-silicate

1. Introduction

Today, materials are as diverse as the components and methods used to assemble them. Fire safety concerns and regulatory issues have fuelled past efforts to use formulated chemical technologies to help protect against unpredictable fire hazards. Flame retardant technology was introduced fifty years ago as bulk 'additive' materials to effectively reduce flammability of naturally combustible, ignitable or inflammable constituents of products used mainly in household products. An example of such materials include polymers which may be easily ignitable. As synthetic materials, their inclusion in industrial and domestic products has generally been considered essential to the safety of end-users either as chemically uncoordinated additives to the target materials or as chemically bonded with textiles [1], coatings [2] or plastics [3] for example during manufacturing. These additives primarily operate by 'retarding' or resisting the ignition phase of flammables thereby delaying their reactive pyrolytic nature to catch fire by insulating against its spread. However, to be effective in retarding flame progression, high quantities of additives

IntechOpen

are required leading to enhanced toxic gas release. They also operate by diminishing flammability when chemically activated or use endothermic processes to drive the suppression of flames. The availability of flame retardants in self-applicable formats such as coatings and sprays has raised concerns on the uncontained exposure of chemicals through inhalation of vapor or through skin contact and other forms of possible contamination. The impact of anti-flame chemicals on the mortality rates [4] has been considerably significant to warrant their widespread use in delaying fire hazards. However in more recent times, health concerns have intensified over the safe use of flame retardants have surfaced— a concern that has emerged more intensified in the era of 'green and environmental chemistry' and authority driven demands for chemical risk to human-health after toxicity related irregularities based material class [5] were highlighted some thirty years ago [6].

There are numerous flame retardant types [7], however halogenated or organohalogen flame retardants containing carbon bonded chlorine (CFRs) or bromine (BFRs) as the major halogen-based constituents are assembled or integrated with polymers forming organic halogenated blends. For example, several flame retardant classes exist [8] as chlorine and bromine polymers while others are composed of phosphorous, nitrogen and sulfur or synergistically made. Others include antimony, aluminum and tin. Since 1992, halogenated flame retardant production has exceeded 20% reaching 25% of the global market and approximately doubling by 1998. The growing demand suggests that flame retardants are a valuable commodity in fire safety and can regulate the combustion process at the heating or decomposition stage, ignition phase and at the level of flame interference curbing its spread by broadly preventing oxygen consumption, heat production and fuel for flame production.

The chemical mode of action occurs at the level of combustion in the gaseous phase through a cooling process that entails curbing the evolution of flammable gases. An alternative action of flame retardants is the deposition of barriers as chemical layers compartmentalizing the supply of oxygen away from the material. Other mechanisms involve additive materials with endothermic properties lowering temperatures below combustible levels or mechanisms that allow the build-up of protective layers cutting off oxygen as essential part of the fueling process. The use of inert materials forming non-combustible gas products also reduce the space for flammable gases entering the ignitable phase. Compositionally, flame retardants are classed by their elementally important counterparts namely bromine, chlorine, phosphorous and nitrogen. The importance of the point of action of flame retardants is shown in **Figure 1** which often depends on the combustion profile of a material and how well it 'fits' with the flame retardant properties of choice [10]. The release of fire hazards are shown in **Figure 2**. In terms of the molecular mechanism of action which are largely unknown, signify only a broad but general based understanding at the bulk scale [11]. It therefore becomes important to re-visit the flame structure partitioned as the outer, middle and inner zones differentiated by color and temperature. Temperatures vary considerably by orders of magnitude in a single flame and temperature variability affects combustibility of materials generating different particle types that originate from different regions in the flame. These regions are further separated by the variable degree of combustibility of vapors dictated by the availability of oxygenated air and critical temperatures needed to ignite unburnt fuels generating high energy propagating particles during flame propagation [12]. It has also been suggested that modeling of flow configuration of flames if adequately understood can help direct particle-particle interactions with multi-scale implications that reside at the gas–solid phase [13]. The contribution of nano- and sub-micrometer particles to flame retardancy is not only dependent on low migration resulting in the surface decomposition of condensed material (char) and the vaporization of particles via surface diffusion at higher temperatures

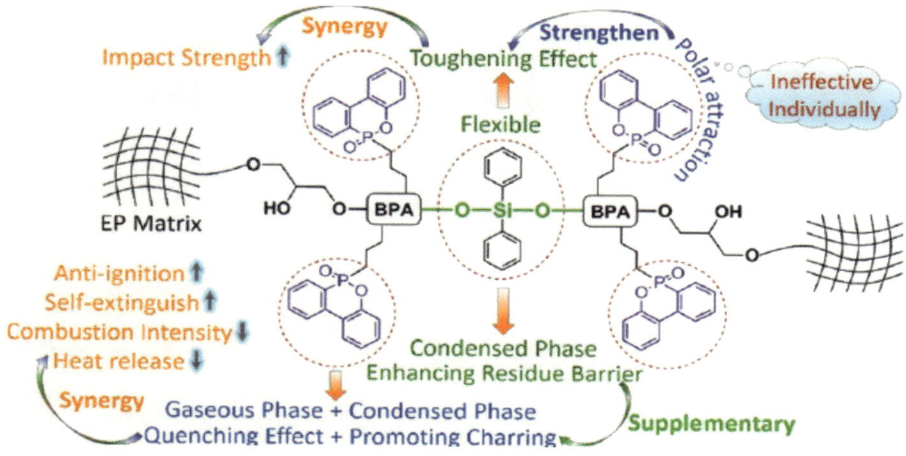

Figure 1.
Flame retardant property and the importance of chemical synergy in the behavior of phosphaphenanthrene retardant chemistry in suppression of flame progression. Reproduced with permission from [9].

phase but also surrounding combustion gaseous conditions. Both particle diameter (geometric mean size) and mass was affected by altered O_2:CO_2 and O_2:N_2:CO_2 ratios during the vapor to particle conversion [14] (**Figure 3**). Given that the pattern of nanoparticle evolution is connected to how a flame is structured, motioned and governed, modeling the underlying features could have important ramifications in reducing or silencing the more turbulent aspects of flames. For example, the size increase of nucleating particles whilst traversing across the flame front rapidly aggregate and particle progression and accumulation can be suppressed by temperature reduction [15].

The basis for the use halogens as flame retardants originates from the production of radicals from the combustion of hydrocarbons such as $H^{\bullet}$, $HO^{\bullet}$ and $O^{\bullet}$ (**Figure 4**). High energy radicals combine with hydrocarbons such as methane to increase radical formation fueling green-house emissions like CO_2 and toxic CO (**Figure 5**) thus considerably contributing to exothermic reactions — a process that is repeated by increasing the chain of radicals during combustion. In particular, hydrogen bromide (HBr) at the ignition phase (decomposes to $Br^{\bullet}$ (bromine radical) and behaves as a quenchers of $HO^{\bullet}$ and $H^{\bullet}$ through protonation forming water and hydrogen respectively. Bromine radicals are able to compete with $HO^{\bullet}$ and $H^{\bullet}$ from hydrocarbons such as methane neutralizing radical hydrocarbon production and instead forming ground state hydrocarbons shown below in **Figure 4**.

Some studies have suggested however that radical recombination although important, constituent CO-H_2 may propagate with differing burning velocities (flame speed progression relative to the unburned state) under conditions of minimal heat loss, particle formation is influenced by the residence time in pre-mixed flames [16].

These chemical events work synergistically in reducing the propagation of flames through flame suppression and delaying flame ignition via chain termination 'type' reactions. The diverse use of combustible materials and their flammable nature has been at the core of safety and protective concerns to humans and the environment. At the forefront of chemical exposure are fire fighters, a risk assessment case study of this vulnerable group has highlighted chemical exposure through combustion as a leading cause for cancer and thyroid complications [17] among the firefighting population from direct exposure and direct inhalation of chemical constituents of flame retardant materials. Tens of thousands of deaths

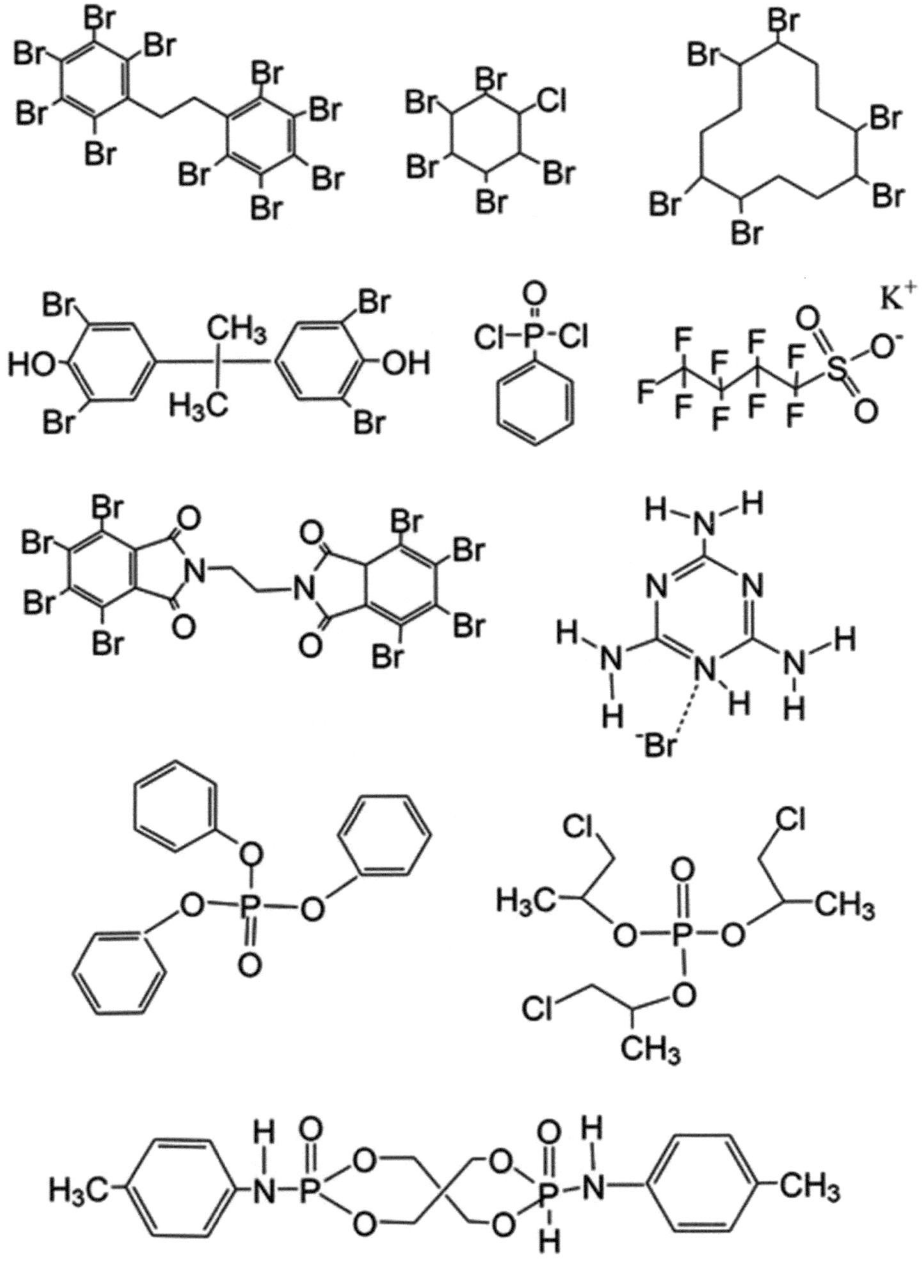

Figure 2.
Halogen derived flame retardants in use.

per year occur through the inhalation of toxic fumes such as carbon monoxide, hydrogen chloride, hydrogen cyanide [18] and dioxins [19] making the effects of smoke toxicity a leading reason for mortality during fire related incidents. The fact that gaseous toxicity release surpasses the ability of flame retardants to suppress fires is a major concern [4] and has shown to persist inside teaching classrooms [20] and college dormitories [21]. Poly-brominated and other halogen-based retardants are shown to be linked to bioaccumulation [22] with child developmental problems effecting child IQ performance, development and intelligence impacting neurologic function in children [23], weakening immune systems through Immunotoxicity [24], reproducibility [25], metabolic [26] and respiratory [27] interference.

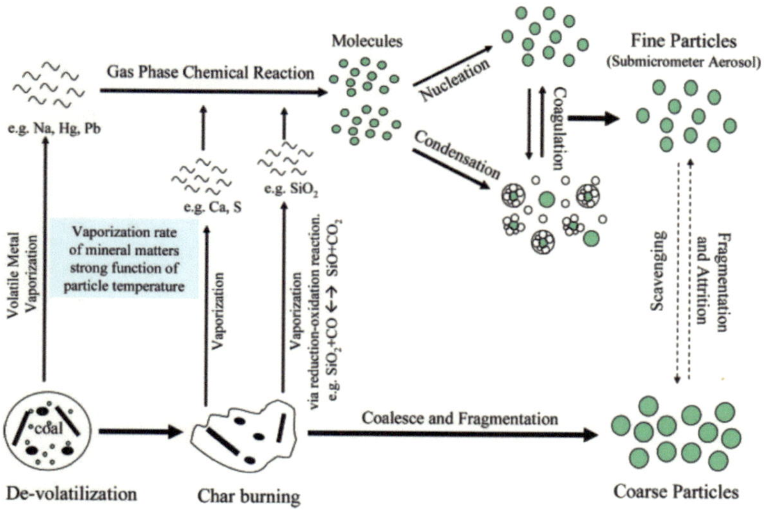

Figure 3.
Condensed to vapor phase particle conversion and associated processes during the combustion. Reproduced with permission from [14].

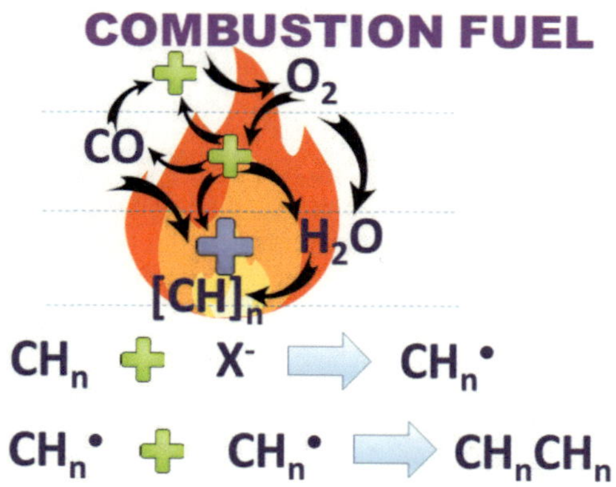

Figure 4.
The global equation of combustion which propagates by a set of chain reactions mediated by high energy radicals that originate from interaction of hydrocarbons with flame retardant halogen anions (Br^-, Cl^- etc.) in a mixture composed of oxygen, carbon dioxide, water and combustion fuels e.g. hydrocarbons.

Such health risks also include harm to wildlife [28]. These chemicals discreetly find their way into the surrounding environments spreadable through the air [29] and water or become attached to dirt, soil, sand and powder particles and subsequently into human contact directly through inhalation or through food and water-intake. These health risks shown in **Figure 6** together with unknown and unsubstantiated toxic effects of flame retardants has caused considerable concern in recent risk assessments [31] fuelling demands for the disuse of constituent chemicals by National Oceanic and Atmospheric Administration and the EU driven Restriction of Hazardous Substances Directive [32]. Environmentalists and health risk advisers have particularly focused on finding alternatives for halogenated flame retardants [33] with a view to replacing them with non-halogenic constituents. However, there exists an 'unmet need [34], to make flame retardants toxic free as new technologies emerge. This situation is being addressed by a drive towards policy changes focused on a class of flame retardant types and steps for their replacement [5]. The current

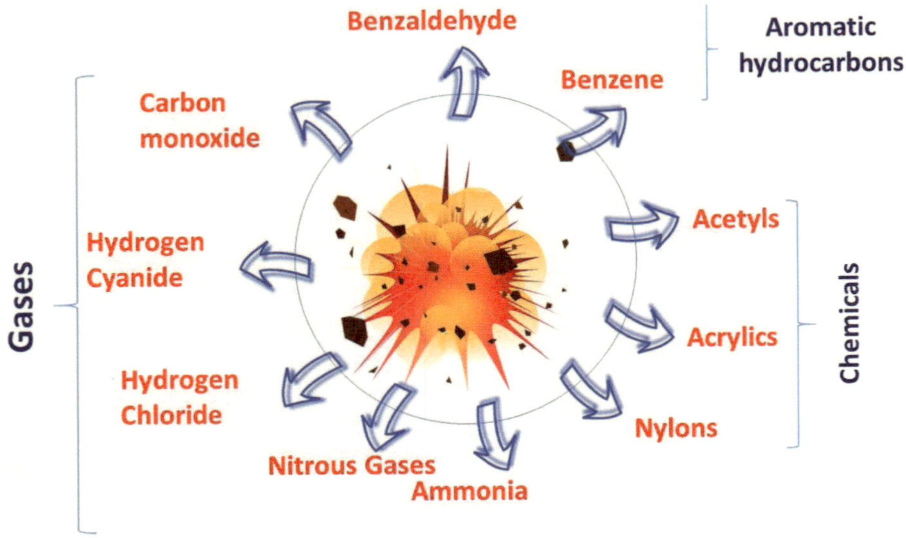

Figure 5.
The release of toxic gases, hydrocarbons and chemicals from polyamide based flame retardants.

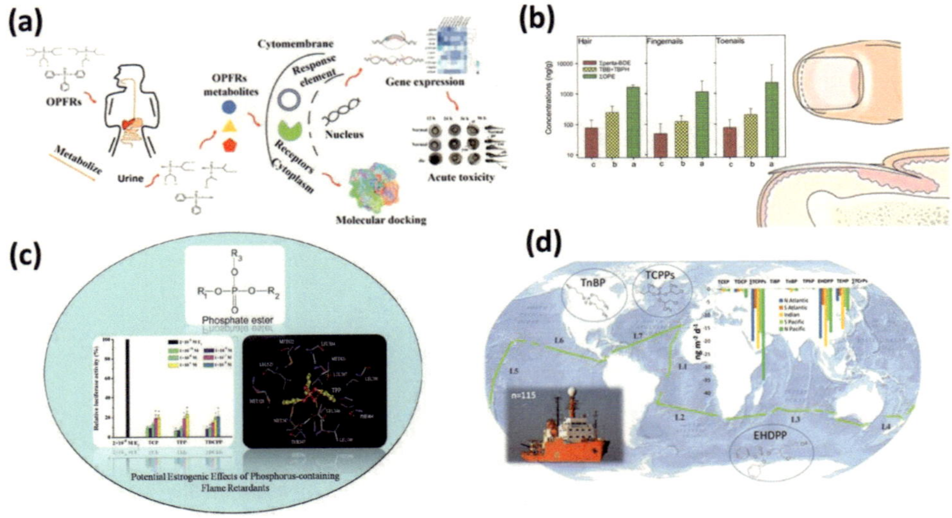

Figure 6.
Prevalence and potential for toxicity of flame retardants (a) metabolic fate of flame retardants affecting gene expression and inducing toxicity (b) accumulation in finger nails and hair follicles (c) hormonal suppression/ activation effects of phosphorous flame retardants modeled through molecule-ligand binding (d) detection of organophosphate esters across tropical and subtropical Atlantic, Pacific, and Indian oceans. The release of toxic gases from polyamide based flame retardants. Reproduced with permission from [30].

challenges weigh on the benefits of flame retardant use in saving lives against death due to toxic inhalation.

In this review, we focus on nanoclay materials and the potential to exploit both their intrinsic and modifiable properties adaptable as flame retardants with a view to make anti-fire materials safer by reducing their potential for toxicity and harm to humans and animals alike. Insight into the technological challenges that confront the flame retardant industry in securing safe and usable chemicals made acceptable by current environmental and health standards is highlighted. The low-dimensional characteristics of nanomaterials is discussed in the context of introducing and exploiting new flame retardant properties which are absent at bulk scales. Particular attention is directed to polymers as high polymer content and their associated toxicities above ignition temperatures are often the cause of fire related fatalities. In particular, the synergistic role of nanoclays with polymer chemistry offer substantial improvements and the potential of these advancements is presented in the context of the underlying challenges that require new explorations to their development.

2. Igniting low dimensional chemistry to extinguish flame propagation

Increasingly, researchers and regulators of flame retardant use are looking towards the materials used by nanotechnology to improve the risks associated with anti-fire materials. Nanomaterial architectures offer considerable structural and functional control as host modifiers in parent structures. Particles of diminished size at nanoscales represent an important class of materials of technological value in controlling chemical and physical complexity. They hold the potential to offer considerable pyrolytic control as a new class of flame retardants through reinforcement of intercalated low dimensional materials. The effects of increased nanoparticle surface area and surface energy at the nanoparticle-polymer interface [35] caters for diverse orientations at nanoscales that impose changes to the overall microscale structure. Favorable polymer-particle associations can result in interactions that the affect physical and chemical properties of flame retardant behavior. Contributions in this respect may arise from surface interactions (in micro or nano confined spaces), in shape formations affecting polymer spacing and those that result from interfacial compatibility at polymer-filler interfaces. The larger elements of the effects that occur at the micro/nano level translate to globalized effects influencing mechanical properties, permeability and flow behavior of gases through interspaces and structures, thermal property and conductivity. Further, nano based flame retardants utilize less chemicals compared to their bulk counterparts diminishing the production of toxic vapors.

3. Current challenges in flame retardant design

The search for new flame retardants has been intense mainly to make their use safer and more effective. The challenge has been to understand more clearly the underlying mechanisms of pyrolysis particularly in the context of polymers to effectively introduce and to deliver new chemistries and modes of action that manifest at the nanoscale. It has been well-known since the 1960's [36] that the liberation of high energy free radicals play an active role in fire propagation and the broad task of 'chemistry' has been to control heat combustion by limiting radical formation. The emphasis has shifted in controlling the decomposition of flame retardants [37] to limit toxic inhalation and the thermal degradation of polymers [38] and to gain a knowledge-based appreciation of mechanisms prevalent in flame propagation. The broader picture here is complicated by flame retardant mechanistic modes and the overlying chemical synergy with material decomposition. This aspect of flame control has been challenging since polymer chemistry is considerably more diverse than flame retardant chemistry making the alliance of synergistic control intellectually and technologically demanding. Vulnerability to fire-spreading scenarios lies within the combustion process itself aided by a number of processes that lead to the ignition phase. Flame retardants while designed to delay combustion and pyrolysis propmote secondary effects of toxic fume emissions and free radical formation from burning materials such as polymers which may override the ability of retardants to sufficiently contain fire progression and toxic smoke. The current problems focus not only on the material chemistry but emphasize a shift towards the intrinsic nature of the material itself. Since flame retardants conventionally operate on the principle of delaying fire progression, current objectives require using new material properties for the implementation of multi-functional flame retardants with design features better suited to the thermal properties of particular polymers. This is emphasized in **Figure 6** which seeks to manufacture a 'new generation' of materials effectively suppressing fuels that contribute to flame production such as oxygen, allow carbon dioxide permeability to reach sites to extinguish the birth of new flames and slow combustion, allow the capture and containment of toxic fumes, reduce the population of high energy radicals through quenching mechanisms and lower temperatures below the ignition phase. Health concerns must also be balanced with the dynamics of environmental issues, performance and costings making flame retardants more easily and economically available. The most attractive direction being pursued are low-dimensional materials that act as fillers for polymers bearing the ability to physically and chemically modify the thermal progression of polymers and other materials and to alter critical factors pertinent to fire control more advantageously with minimal damage to the surrounding environment. The additive or synergistic role of nanoclays hold much promise in this direction.

4. Polymers - a fire safety perspective

Polymers of varying compositional structures are wide spread in many everyday items [39] both in synthetic and to a lesser extent in biomaterials are a probable source of toxic inhalation during decomposition in fires. Most constitute a fire risk in many household appliances, furniture and wearable materials as the elements of polymer combustion drive fire propagation from the point of ignition. A growing body of evidence suggests that the reinforcement of polymers show a strong correlation with fire resistance while reducing the need for excessive use of fire retardant as single component materials in goods. The physical and chemical relationships between polymer behavior and fire progression has a unique place in engineering with the potential to offer new flame retardants that are both safe and effective. The complexity in the development of effective flame retardants is depicted in **Figure 7** [40] and empathizes the potential for multi-levels of combustibility that lie between the condensed (liquid) and expanded (gaseous) interfaces. As pointed out by Huiqing [40], the combustion cycle of polymer materials subject to a pyrolytic state pass through an [1] initial heating phase, reaching the [2] decomposition phase followed by the [3] ignition phase. Lastly, the combustion phase is triggered by the volatility of the liquid or gaseous products of decomposition at the ignition temperature and the excess heat generated via combustion in the presence of O_2 is utilized at the polymer surface to release more combustible products. This cycle contributes to increasing the free radical population and further increased chain branch reactions as described earlier in **Figure 4** accelerating the cycle of flame growth. Knowledge of the molecular properties of parameters that can be controlled is (shown in the orange box in **Figure 8**) can be applied and incorporated as design features in flame retardants to enable more effective fire suppression. Important targets in flame retardant design are effective suppressors of combustible compounds and their rate of evolution in keeping ignition temperatures below combustible thresholds. Some strategies that have been described include water formation through the use of brominated flame retardants [41] to contain temperature elevation via free radicals operating as inhibitors of flame propagation elements. Other forms of flame resistance includes charring [42] or partial burning

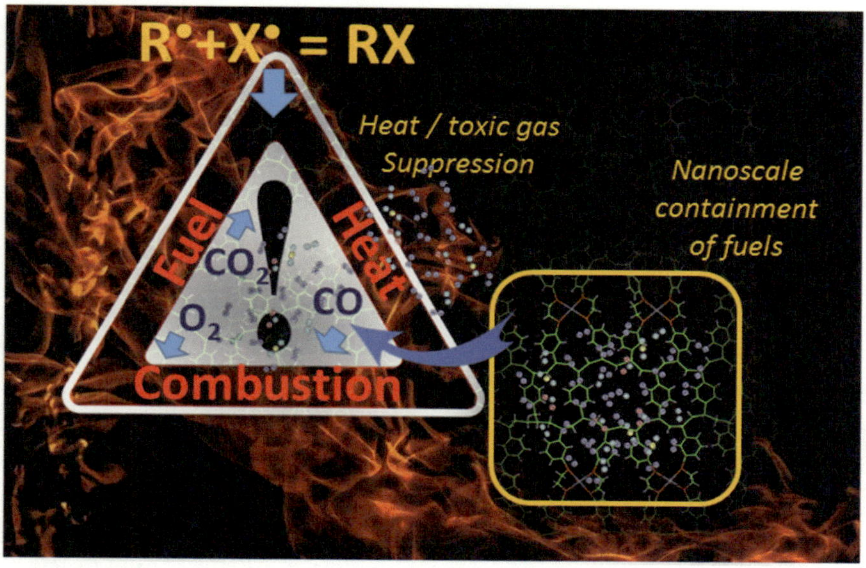

Figure 7.
Low-dimensional materials with multifunctional properties to combat fire hazards on multi-levels.

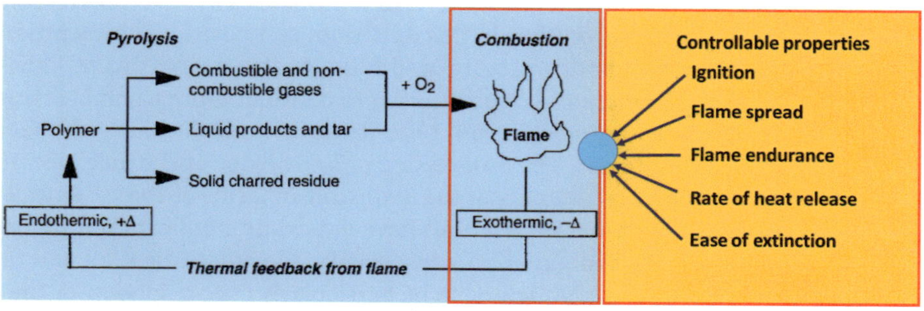

Figure 8.
A schematic showing a general role of polymer pyrolysis in combustion at the gas, liquid and solid interface and the potential for introducing key design features in flame retardants guided by key controllable properties. Modified with permission from [40].

which reduces polymers to black carbon essentially removing oxygen and hydrogen and effectively masking polymer surfaces (as assessed by measuring the limiting oxygen indexes or LOI's) [43] from further combustion through oxygen fuelling and volatile vapor mixing. In the oxidized form, char forms nitrogen oxide and heterocyclic compounds providing a basis for chemical fuel production at higher temperatures. Using two coal types as a carbon based combustion model, peak heat release correlated with the time to ignition showed the char content to decrease or increase with particle size increase [44] of the coal type investigated. Char forming chemistry has been investigated [45] and viewed as a useful alternative to halogen free retardants for polymers exhibiting fire resistance. The widespread application of organic and synthetic polymers in the manufacturing sector and related technologies increases the potential for toxic gas release by decomposition in case of fire demands safer designs of manufacturing protocols and newly improved fire resistant coating materials to replace existing ones. While char formation shows good potential to act as a shield preventing oxygen from mixing with polymers, char chemistry still remains a potential source for flammable gas release under persisting temperatures (**Figure 9**).

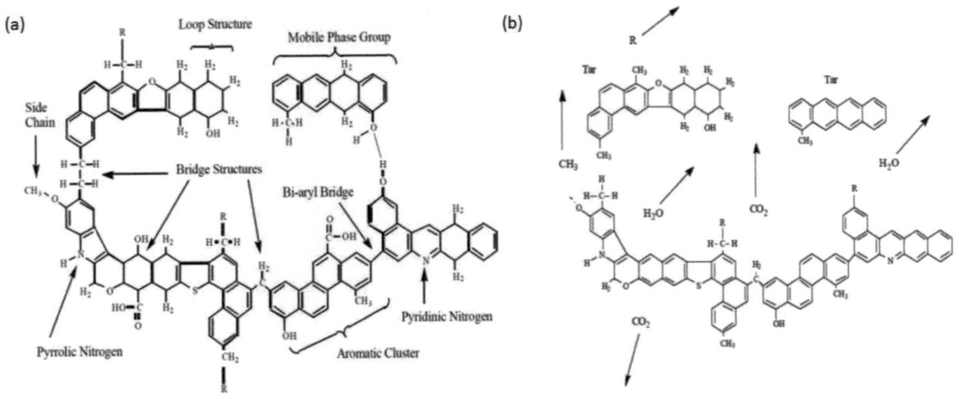

Figure 9.
(a) Likely compositional elements of char (coal) representing (b) possible sources of volatile chemical fuels at ignition temperatures. Reproduced with permission from [46] and modified from [47].

The hazards associated with fire retardant materials arise as a result of fire and the onset of flammable products released could be abated by slowing the decomposition rate of temperature polymer disintegration. Re-tuning the thermal behavioral properties of the primary polymer with other materials as additives that intervene with key elements that regulate combustion related properties to scale-down flame propagation is an important objective. In this direction, compositional integration with nanofillers can provide both structural and functional elements to alter the thermal properties of polymers and opportunities to consider more palatable flame retardants or to introduce new mechanisms to control and limit the harmful effects of existing ones currently in use. The progression of knowledge of the mechanical effects of particles on polymer stability at the nanoscale in terms of the structural harmony between filler-polymer interactions have important implications but have rarely been discussed in the context of polymer matrix, the nanoscale filler and the interfacial region [48]. This is highlighted to be particularly crucial in view of the ability of nanoparticle networks to diminish the combustibility of polymers [49].

Reinforcement of polymer strength and rigidity through stiffening uni-dimensionally or multi-dimensionally by pushing the chemical equilibrium towards char formation and increasing the barrier properties of chars could lead to new methods for fire retardancy. Hence, efforts have been directed for enabling mechanical stiffening of polymers [50] in a variety of ways that affect the flammability properties [51] closely coupled to ignitability [52] leading to a more controlled combustion arrest phase. For example, addition of co-monomers such as acrylonitrile, butadiene, and methyl methacrylate to the pyrolytic properties of polystyrene and styrene [53] and 2D-MoS_2 nanosheet-containing polyurethane [54] have been fabricated. However, more recently Varol et al. [55] determined that nanoparticle nanofiller quantity and not size determine the strain hardening in polymer nanocomposites (**Figure 10**). Although this study signifies mechanical strength in the context of load bearing application of polymers, tuning strain hardening behavior with nanofillers with a reinforcement magnitude that allows the decomposition of polymers to be critically restrained at high temperatures has much value for tailoring flame retardants. Replacing the characteristics of brittlement of polymeric materials with heat resistant materials [56] by enhancing higher temperature performance in nanocomposites has meritable arguments for the flame retardant industry. A key challenge in the development of environmentally acceptable and low-to-no health risk flame retardant additives for use with conventional anti-fire hazard materials has accelerated in recent years. In line with the growing importance of nanomaterial

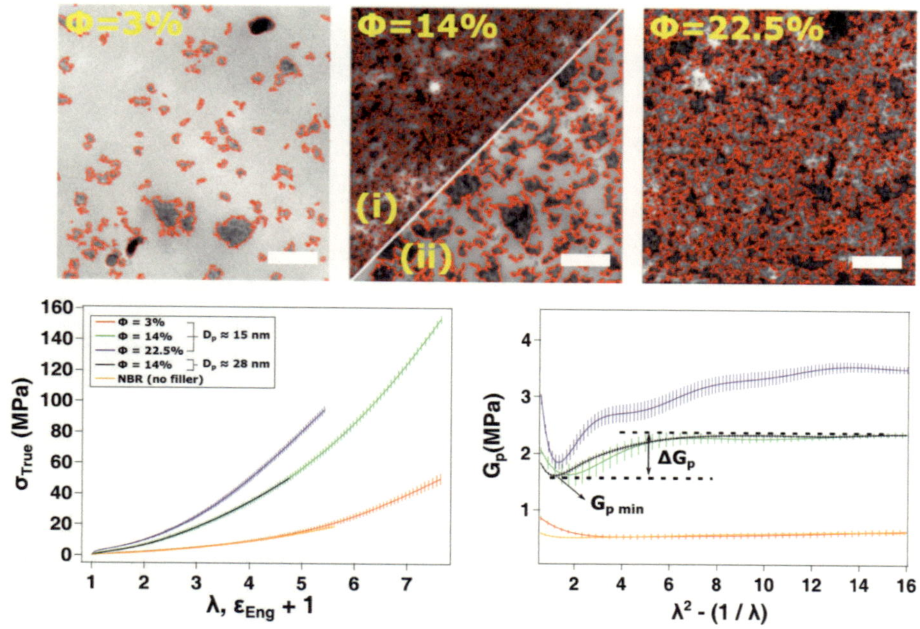

Figure 10.
(Top panel) Transmission electron microscopy images of the volume fraction (Φ) of SiO_2 derived nanocomposites particle size with respect to particle size. (Bottom panel) Plot of the true stress (σ) as a function of the extension ratio (λ) for particles exhibiting different filler volumes and size. Modified with permission from [55].

components for use in flame retardant formulations, the focus of interest has increasingly shifted towards clay based-minerals as nanofillers [57] to address some of the key challenges discussed. Multi-layering of flame retardants in nanocomposites could be an attractive structural feature to allow other additives to interact with flammable constituents that may persist even after part thermal stabilization of polymers with nanoclays. In addition, intercalation between polymer and nanoclay layers offer a level of intervention at the molecular nanoscale scale that may operate synergistically with the polymer and flame retardant chemicals. The possibility of introducing barrier properties, mechanical strength of polymers and hence diminishing volatile decompositions and support flame retardant properties cooperatively.

5. Nanoclays composites: challenging the role of conventional flame retardant mode of action

Nanoclays form a class of inorganic clay based nanomaterials (**Figure 11**) with chemical and structural attributes that enable their integration with diverse materials as clay nanocomposites including polymers [59]. They exist as silicate/aluminum-silicate structures in the form of montmorillonite, bentonite, kaolinite, hectorite, and halloysite. Nanoclays comprise layers of 1 nm thickness separated by interlayer distances between 70 to 150 nm modifiable as nanocomposites through intercalation with guest structures. While compositional mergers of nanoclay result in superior mechanical and tensile strength, properties aligned to reduce gas permeability [60] is achieved through the deposition of thin coated layers, alterations in glass temperature (temperature distortion) of nanocomposites and changes in modulus occur proportionally with increasing amounts of nanoclays and significantly alter material characteristics. Further nanoclays could be attractive additives as anti- combustion materials due to the differential permeability

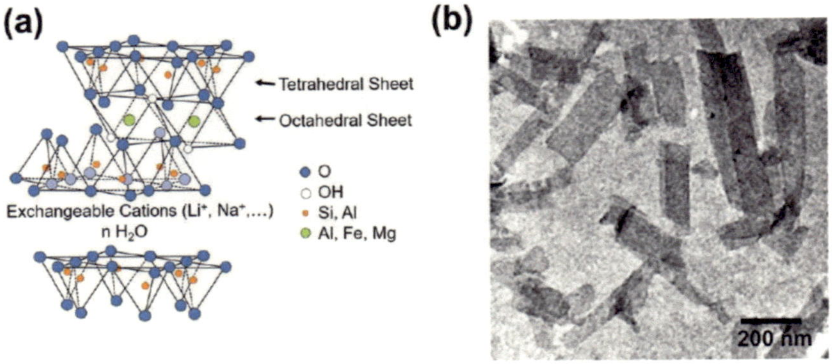

Figure 11.
(a) Structure of nanoclay exemplified by the (b) TEM image of nontronite. Reproduced with permission from [58].

behavior to oxygen and carbon dioxide. Nanoclay exhibits better gas barrier properties against oxygen then carbon dioxide and thus has the potential to limit oxygen as a fuel source to flames [61]. The possible tuning of the molecular spacing of silicon tetrahedral units suggests that gas permeability can be altered to a particular gas through morphological changes to the filler as in the case of nitrogen [60]. With special interest to polymer based flame retardants, the dispersion behavior and orientation properties of nanoclay particles can find significant use in generating intercalated layers with polymers that largely depend on their molecular assembly and ability to interact [62]. Also in the interest of minimizing the release of toxic fumes, nanoclays can be used to generate effective flame resistant barriers by making use of nanometer scale surface adhesion properties that extend across micrometer lengths with aspect ratios between 200 ~ 1000 or indeed higher [63]. Hence a strong correlation of barrier properties to geometric considerations such as orientation, filler morphology and aspect ratios are critical to their functional use as additives in flame retardants to control the spread of fire [64].

To advance fire retardant chemistry however, there exits an underlying need to understand better the molecular transition from the homogenous nanoclay and polymer architectural state to a structurally unified polymer-nanoclay nanocomposite state at the nanometer level. An immiscible state between nanoclays and the surrounding polymer environment lacking a chemical union of bonds as a co-mixture rather than a self-ordered intermixture does not allow ascendency to a superior consolidated mechanically enhanced state. In other words, the filler chemistry must adopt bond, shape, physical and chemical structural characteristics unique to the assembly not shared by the components separately or as a co-mixture. In order to allow the next generation of fire retardants to be of greater general applicability and of broader scope to different polymer populations, the exfoliation of nanoclay inter-layers requires the use of interactants (modifiers such as inorganic minerals, synthetics, hydrophilic or hydrophobic solvents, dispersants, reductants/oxidants and other chemical agents under conducive physical conditions) so that the chemistry and spacing between interlayers is compatible with the polymer environment.

6. Synergistic effects of nanoclays

A well-recognized route for the suppression of flames and probably the best studied process has been described by carbonaceous char formation exhibiting

Diameter nm	70 nm	80 nm	90 nm	100 nm	110 nm	120 nm	130 nm	Total	Mean Diameter nm
Volume %	7.58	12.12	15.15	15.15	15.91	14.3	19.7	100%	99.29

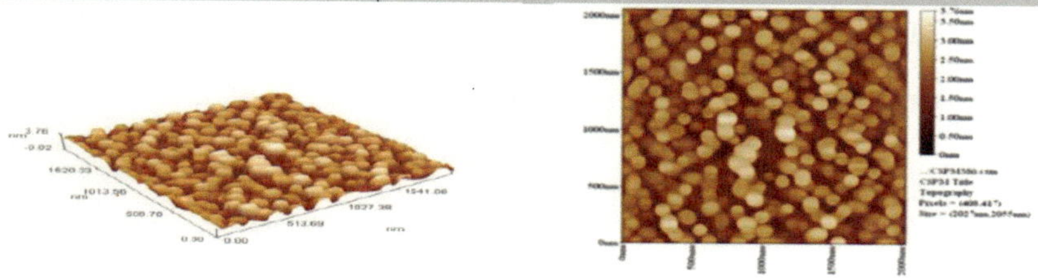

Figure 12.
Layered geometry visualization of nanoclay by atomic force microscopy showing the importance of ionic interpenetration at the nanoscale with particle diameter changes from 70–130 nm. Reproduced from [67].

barrier properties [65] that effectively 'cut-off' gaseous fuel supply reducing flame capability. Decade old studies on nanoclay composites have shown to provide 'multistage' insulation during fire progression, however the dependence of fire performance on a measurable quantity of nanoclay was prominent ranging from optimal to infective in delaying and an advancing fire hazard [66]. Such studies have highlighted the complex chemical nature of nanoparticle interpenetration between layers (**Figure 12**) at nanoclay interfaces and the limited knowledge in selecting and activating chemically useful routes that diminish flame growth against accelerating factors. The ambiguity in the nature of these mechanisms is complicated by molecular changes that occur with nanoclays at the surface of different polymers pointing to the nature of interacting species that may be consumed or liberated during different stages of thermal reaction. Char formation is mediated via the catalytic crosslinking of polymers with properties that are identifiably different to their constituent reactants both by their physical and chemical nature. Nitrogen metal complexes have been suggested to counteract barrier char formation by catalytically weakening polymer stability through site specific decomposition by virtue of the metallic catalytic sites accommodated within the clay region [68]. The extent to which these processes operate play a central role in determining polymer stability and the dominant mechanisms that ultimately prevail However, dispersion becomes an important criteria at the nanoscale level in determining effective flame retardant properties [69] while a more informed selection of polymer types forming the nanoclay-composite chemistry can used to increase the carbon content of char [70] likely favoring crosslinking ability and subsequent barrier properties. The ionic nature of nanoclays also plays a significant role in governing the flame retardant properties of complexes.

Polymer variability in terms of composition, structure and intrinsic properties are defining features that determine the degree of material inflammability. Understanding the mechanistic role of nanoclay particles in subduing polymer vulnerability to thermal heating and volatility becomes a challenging problem in the field of flame retardancy. **Figure 13** shows the chemical integration of polymer-nanoclay surface where the interface chemistry is a key factor in deciding flame retardant properties of the resulting nanocomposite. Some generalized insights into possible mechanisms have emerged from fire retardant investigations. As mentioned, much effort has been directed in uncovering the complexities surrounding char deposition pathways to improve mechanisms that have an obvious advantage in blockading the mass transport of fire enhancers (e.g. free radicals, gases,

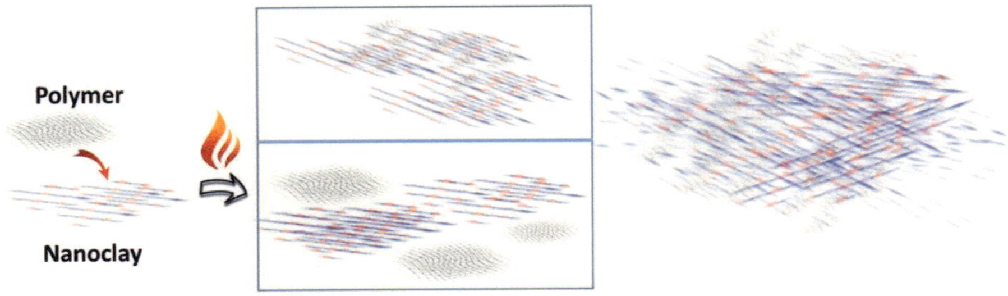

Figure 13.
Clay-polymer assembly processes under thermally elevated conditions determine fire retardant potency driven by nanoscale chemistry.

hydrocarbons etc.) that can be released from clay surfaces via an initial pyrolytic state. Barrier properties become progressively more effective during the condensed phase as more material is retained and cross-linked leading to enhancement with subsequent layers. The synergistic cooperation between nanoclay layers and mineral addictive's has attracted much attention in recent years and has rapidly become the most favored approach in the field of flame retardancy.

While the immobilization of decomposed components aided by low migration at the surface of silicate layers is an important mechanistic approach to barrier creation and polymer modification, the unrestricted migration of the more volatile degradation products can release free radicals with the ability to passivate halogen radicals through the chemical association with hydrocarbon radicals. Such radicals may take the form of metals embedded or chemically attached to the nanoclay polymer matrix. The intrinsic and selective nature of the polymer materials modifiable by thermal processes potentially embody exploitable synergistic characteristics particularly at the nanoscale. For example, improvement in interfacial adhesion has been correlated to a change in thermal decomposition to elevated temperatures during the merger of nanoclays with flame retardant supporting the synergism of nanoclays to char layer enhancement [71]. Hence, the chemical synergy between the reactive nature of starting materials (nanoclay, polymers and minerals), the initiation temperature and pathway for thermal activation of condensed and volatile degradation products and the knowledge of key mechanistic events in terms of mapping structure to flame retardant function and performance still need to be resolved. Certainly, the correlation between clay composition and decompsition kinetics ohas been identified to be a key factor in charring [72]. Despite the hidden challenges which persist in the field, the effects of scale are slowly emerging using a number of experimental approaches. Some observations that form the basis of well-accepted outcomes coupled to unresolved mechanistic ideas are generalized in **Figure 14**.

The potential for polymer-layered nanoclay composites for applications as flame retardants has been recognized in earlier studies [73]. A common objective in nanocomposite design has been the search for effective interacting agents for the alignment of polymers along the corridor-like arrangement in stacked layers of nanoclays to improve mechanical strength, thermal stability and gas barrier properties. Diagrammatically, the stacking is described as intercalated in which polymers lie between inorganic layers in well-ordered arrangements or adopt exfoliated patterns represented by disordered arrangements within the layers exemplified by poly(ethylene oxide) adsorption on nanoclay surfaces (**Figure 15**). In similar studies [75], the use of polypropylene-graft-maleic anhydride and polystyrene-layered-silicate nanocomposites in their ability to suppress flames was mechanistically similar and it was established that the behavioral properties depended on the degree

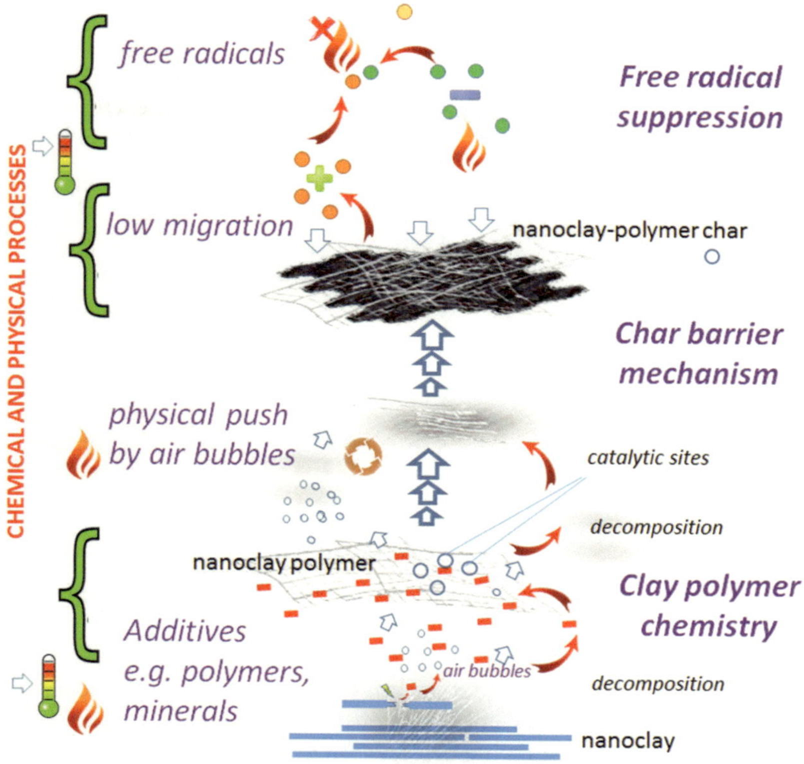

Figure 14.
General mechanistic routes to fame retardancy by char barrier formation and free radical suppression mediated by catalytically induced sites at the nanoclay polymer interfaces. Migration of particles formed by degradation products originating from pyrolytic chemical processes are assisted by air bubble movement pushing particles upwards and low migration of condensed phased char particles.

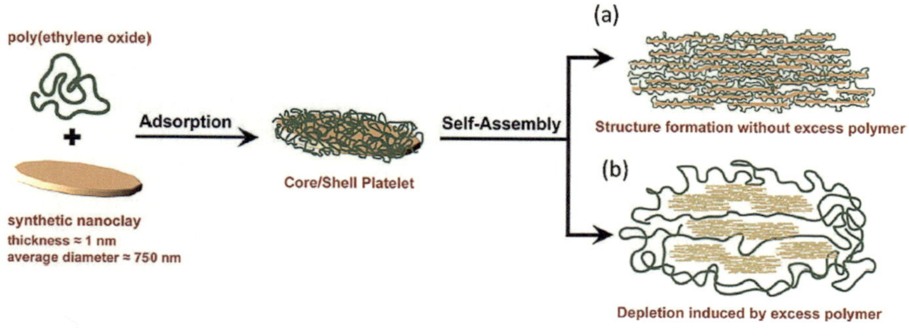

Figure 15.
Schematic showing two types of silicate-polymer composite layering either as (a) ordered (intercalated) or (b) disordered (delaminated) arrangements. Reproduced with permission from [74].

of dispersibility and type of silicate material. The effects of flammability reduction were inferred through a protective insulation surface slowing down the decomposition rate of the underlying material and thus preventing exposure to the decomposition products. Mechanistic insight has been better aided by knowledge of the molecular orientation of nanoclays that have resulted in establishing three types of nanoclay geometries. Understanding the geometrical orientations of nanoclay shown in **Figure 16** may help in our decipher the barrier properties exhibited by these class of materials which exist as (a) phase separated (immiscible), (b) intercalated or (c)

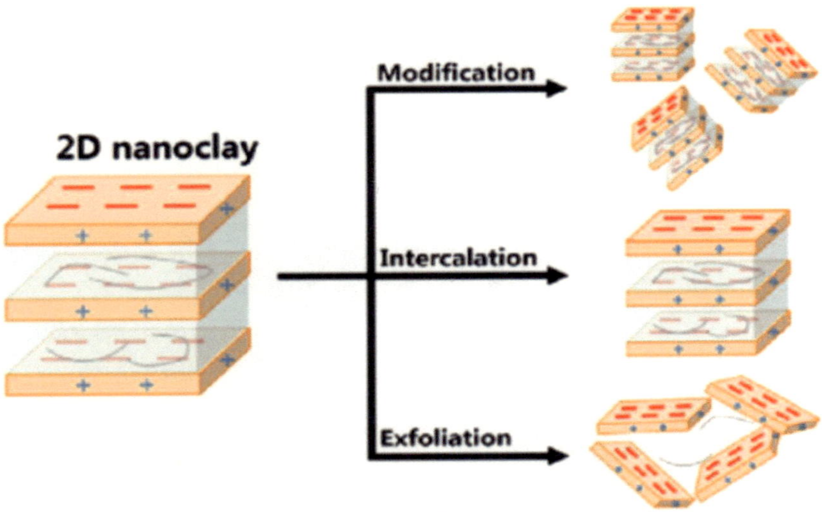

Figure 16.
Differing morphological states of polymer–nanoclay structure of nanoclay. Reproduced with permission from [76].

exfoliated as polymer nanocomposites. In (a), polymers do not reside inside the inorganic layers that may exhibit conventional silicate layer properties as layered blocks while (b) shows interlayer distancing of a few nm between silicate layers which is a result of polymer intervention within the layers by the surrounding polymer. Finally (c), shows an exfoliated structure where clay layers are singularly separated by the polymer. Polymer matrix and nanofiller interactions play an important role for atomic-scale nanoclay dynamics. The interaction of the cationic surface of nanoclays with organic layers of polymers is identifiably important in the advancement of flame retardant nanotechnology. There are many reported studies that have pursued the effects of polymer matrix/nanofiller (nanoclay) associations and while there are benefits to the formulations, the precise role of nanoparticle ambulation during surface modification remains unknown. Current knowledge however, stipulates that organic components of polymers can sufficiently reduce excess energy (surface energy) [77] by minimizing cohesive and adhesive forces at the solid-wet interface between cationic and polymer states of matter respectively while supporting the stabilization of interlayer spacing. Spacing serves a useful purpose in reducing surface energy and preventing nanoparticle aggregation generating enhanced dispersion and the level of ordered intercalation largely dictate the nanocomposite structure. In context to **Figure 12**, it might be expected that separation achieved in exfoliated nanocomposites generate maximum space and hence maximum surface area allowing polymer migration to occur and spread along interfaces unperturbed between silicate layers. Schartel et al. [78] demonstrated differing morphologies of phosphonium-modified layered silicate epoxy resin nanocomposites but structures closer to exfoliated forms (**Figure 16**) with silicate as a nanofiller revealed a barrier effect but had little effect on the suppression of decomposition products and combustion of volatiles. The current perception of exfoliated surfaces yielding superior surface and mechanical properties is a phenomenon that is now less clear and conflicting evidenced by experimental interpretations. By its nature, exfoliation generates better dispersity between the nanoclay surface in an uninterrupted polymer matrix. Variability in experimental observations often uncovered inconsistency from glass transition values and suggest that a state of higher thermal stability depends on other factors and not solely on the visible state of a structure. Hence, the level of unpredictability associated with well-separated exfoliated layers of separation could reside with the degree of variability in interfacial patterns that may ultimately determine polymer mechanics and further complexities introduced by compatibilizers as components for in epoxy adhesives [79] for example, or as thermally reactive additives. The lack of understanding of these factors could be contributing factors to discrepancies among related samples and more insightful explanations are needed to develop favorable material properties.

The use of non-halogenated flame retardant materials have recently come into effect like aluminum diethylphosphinate (ADP) [80] improving char yield. It has been reported that the multi-phase use of ADP in the vaporized and condensed phase [81] significantly contributes to flame inhibition. Investigations by Kaynak and Polat [82] further established the role of path driven intercalated/exfoliated NC layers that contributed to the emergence of insulative barrier properties as judged by the increased limiting oxygen index. Chemical synergy between nanoclay layers and ADP likely favored the retainment of aluminum phosphinate at the clay surface forging the incorporation of the mineral within the layers of the carbonized char residues within the mechanically strengthened polymer structure of diminished chain flexibility. Thermally activated vapourization of ADP triggered the oxidation of surface unbound minerals in the gaseous phase to phosphinate radicals HPO2•, PO•, PO2• creating opportunities to neutralize the toxic effects of H• and OH• curbing flame growth. Here, the use of 5% silicate layers was effective in combination with 15% ADP.

In an attempt to better elucidate the mechanistic impact of nanoparticles during pyrolysis, two-dimensional transient-state models based on nanoclay layering using carbon nanotubes as potential polymer reinforcement additives. The study revealed a number of nanoscale effects between charred and uncharred regions that span the clay network serving as the major interface in regulating the transportation of surface degraded and diffusive particles that affect thermal transfer [83]. However, the synergistic shift in surface chemistry at the clay-nanoparticle that permits mass transfer loss via polymer degradation by addition of carbon nanotubes during the initial stages of pyrolysis suggests that improved stability around structured nanoclay layers will be pivotal to controlling flame retardant properties in future nanocomposite designs. Tuning the thermal stability of nanoclays with compatibilizers can provide access to the incorporation of smarter materials such nanopolymer confined quantum sized TiO2 [84] that demands better exfoliation of silicate layers improving immiscibility of layers with TiO2 [85].

Suter et al. [86] have more recently used a novel multiscale modeling to shed light on the mechanisms driving exfoliation behavioral properties of clay-polymer nanocomposites. The results show how molecular simulation techniques targeted around clay interfaces using free energy profiles and structure based coarse-grained iteration processes to understand clay layer exfoliation and how such interactions lead to experimentally observed changes. This was achieved by simulating the interactions of montmorillonite clay, a polymer (PEG) and a quaternary ammonium dimethyldioctadecylammonium ionic surfactant. **Figure 17(a)** shows molecular dynamics simulations of models predicted I, II and III after a simulation of 3 μs with model II and III forming partially and fully exfoliated single layers respectively. **Figure 17(b)** shows the radial distribution as a function of sheet distance and provides evidence supporting a completely dispersed system. In the bottom panel of **Figure 13**, different stages of the molecular dynamic simulation of model II is presented illustrating (a) the clay structure prior to the interactive state with PEG (b) intercalated nanoclay-PEG formation resulting in increased gallery separation and the (c) departing motion of neighboring of layers favoring an exfoliated increases in Young's modulus derived from the stress–strain behavior models of II and III. The

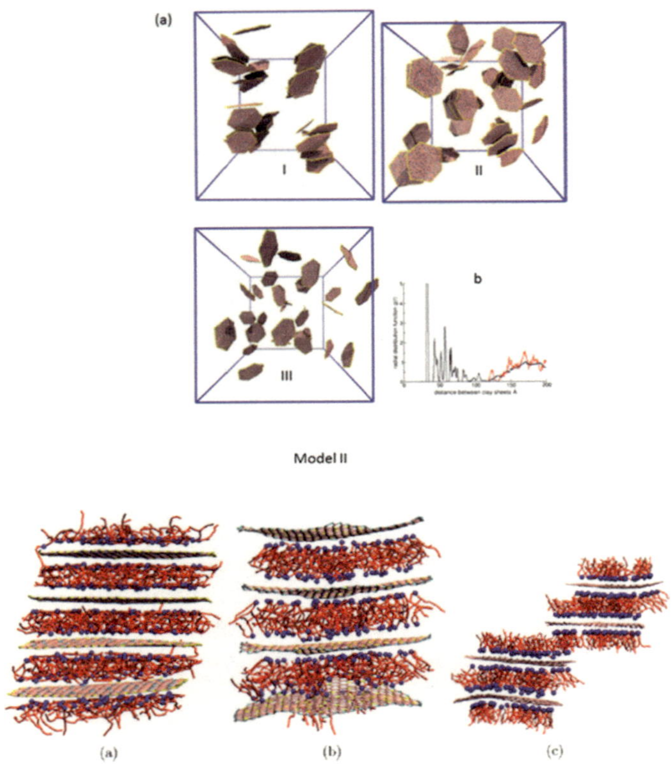

Figure 17.
Dynamic simulation models of (a and b) of (c) partially and fully exfoliated nanocomposite state. The exfoliated models are found to exhibit considerable elastic properties with substantial structures based on model II as described in the text. Modified with permission from [86].

density charge increase around the polymers attributed to the surfactant molecules rendering the nanoclay structure to an exfoliated state by inhibiting a diffusive state which is assisted by a transversally sliding motion of clay. Such simulations provide plausible clarity on the dynamics of nanoclay sheets and important mechanistic clues that can be evidenced and supported by experimental findings with additives. As simulation tools become more powerful, it will become easier to assign definitive structure–function relationships to earlier observations that have been more phenomenological in their explanatory findings. The re-organization of intercalated or exfoliated states in nanoclay based nanocomposites is a promising way to better regulate not only polymer behavior but also the behavior of flame retardants under nanoconfinement signifying the functional role of clay particle size [87].

Nanoconfinement of polymer chains in nanoclays can influence polymer behavior affecting the chemical and physical behavior with additives in the melt, altering decomposition, thermal degradation and intermolecular interactions kinetically of components globally determined by the local morphological states that adapt to intercalated or exfoliated or indeed partial intermediate states. Over the last two-to-three decades, polymer research has intensified the use of nanofiller materials to form polymer bends into tunable nanocomposites. Nanofillers in the form of layered silicates have enabled a considerable reduction in the use of conventional load bearing modifiers in comparison to their low dimensional counterparts by as much as $3 \sim 8$ fold of the total content normally used by conventional processes. A significant trend that continues to be a promising pathway for the future use of flame retardants is the chemical synergy driving nanocomposite assembly while extinguishing the properties of its individual components in the blend. Some of these examples include the synergistic effects of clay-organic intumescent hybrid systems. Distances among silicate layers in nanoclay modified with organic surfactants show functional significance in the absence and presence of intumescent agents. Irrespective of whether nanoclays adopt an exfoliated or intercalated state of matter, increase in the layer distance from 9.8 to 13.8 Å revealed that the stacking morphological architecture of the nanocomposite was important to its role in reducing heat release rate through the formation of dense char layer [88].

Here, we provide a summary of some of the more recent approaches and strategies in flame retardant design used for targeting flammable plasticizers widely used industrial polymers with reference to some key studies on nanoclay-polymer nanocomposites from pervious works. In earlier work, quin et al. [89] in their investigation with nanoclay-polypropylene aimed to clarify some unresolved aspects of key mechanisms of flame retardancy. The fundamental problem related to the dispersion state of the nanocomposite and addressed its influence on flammability which tied in with the underlying role of nanoclay in thermal oxidative degradation and combustion. It was concluded the delay to combustible ignition of volatiles was in fact due to the char barrier formation and the route to char deposition occurred via catalytic dehydrogenation and crosslinking of the nanocomposite which was mainly assisted by acid sites of the silicate. The process is summarized in **Figure 18**. An important component in this mechanism that is most relevant to the liberation and availability of silicate catalytic sites was the addition of the alkyammonium salt which in itself decomposes leading to the pyrolytic release of ammonia and olefin hydrocarbons and the accompaniment of rich acidic cations on silicate surfaces. While this mechanism was effective in suppressing flammability, evidence continues to accumulate and support increased mechanical strength and hardness capacity of nanoclay modified polymers enhanced by significant increases in tensile strength and stiffness using tetramethylammonium chloride but may decline with suboptimal quantities [90]. For example, the addition of nanoclays have shown to disadvantage the working performance of flame retardants as noted for a phosphorous-based retardant [91] comprising a Diglycidyl Ether of Bisphenol A (DGEBA) resin composite. This emphasizes the importance of tailoring the cooperativity between the filler and polymer space and dispersion which facilitates the nanoscale chemistry of the retardant components in the composite in response to parametric changes affecting interactions at the molecular level.

A future direction in the controlled degradation of polymers using regulated by filler amounts is the prediction of degradation products. In a recent investigation,

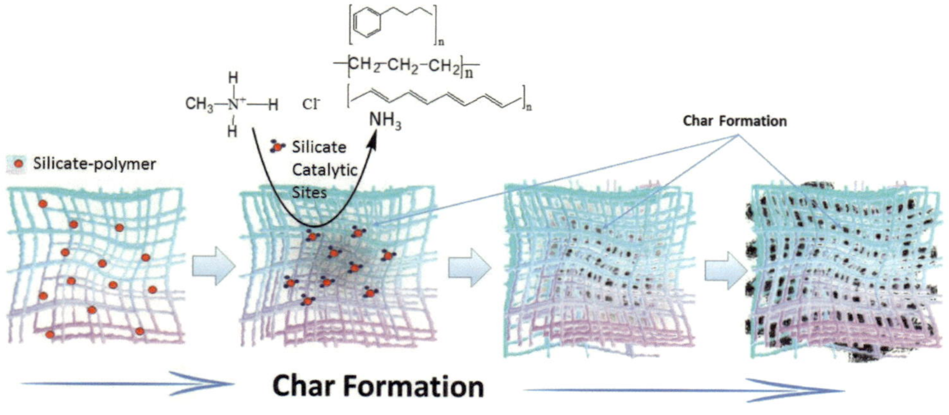

Figure 18.
Active-site catalytically induced char formation during combustion on clay surfaces supported by polypropylene as the polymer matrix.

Saha et al. [92] applied reactive force field molecular dynamics simulation to polyacrylester in which clay nanofillers in conjunction with graphene was used to modify the elastomer polymer characteristics. The model developed using this approach very reliably predicted the effect of adding graphene oxide (GO) in relation to the evolution of degradation products which agreed well with the experimental outcome. Improving the knowledge base of volatile materials originating from different polymer compositions computationally will certainly have meritable consequences in implementing better control strategies against flame enhancing volatile materials [93] and predicting combustion behavior such as ignition times [94], char oxidation and particle sizes [95]. The differences in melting and shrinking behavior that arise from physical conditions affecting ignition performance for example and other important parameters may not be easily accurately measured by micro-scale tests [96] will significantly shift the burden for reliability more towards computation methods.

7. Concluding comments

The current perspective on nanoparticle driven flame retardant performance specific to nanoclay materials emphasizes their growing importance as fire regulatory materials and reduced toxicity. At the nanoscale, clay interactions with different polymer types hold the potential to steer nano-driven mechanisms to more effective outcomes in retarding the spread of flames. In the generality of flame retardant mechanisms to the formation of chemical barriers and passivation of fire enhancers such as free radicals, nanoclay makes important contributions to polymer structure and mechanics, catalysis and kinetics of production degradation and acceleration of stability of factors in slowing down flames. The thermal activation of new catalytic sites at the clay-polymer interfaces will crucial be in developing a more effective class of retardants in the future particularly in the context of target specific modifiers. The lack of detailed mechanistic knowledge of nanocomposites is often hindered by the limitation of experimental approaches and computational approaches will become important predictive tools for guiding areas of new insight.

Author details

Sanjiv Sonkaria and Hyun Joong Kim*
Lab. Adhesion and Bio-Composites, Major in Environmental Materials Science, Department of Agriculture, Forestry and Bioresources, Center for Adhesion Research and Development (Smart CARD), Seoul National University, Seoul, Republic of Korea

*Address all correspondence to: hjokim@snu.ac.kr

References

[1] (a) Bourbigot, S., 2 - Flame retardancy of textiles: new approaches. In *Advances in Fire Retardant Materials*, Horrocks, A. R.; Price, D., Eds. Woodhead Publishing: 2008; pp 9-40; (b) Horrocks, A. R.; Kandola, B. K.; Davies, P. J.; Zhang, S.; Padbury, S. A., Developments in flame retardant textiles – a review. *Polymer Degradation and Stability* **2005,** *88* (1), 3-12; (c) Flame Retardants: Textiles. In *Van Nostrand's Scientific Encyclopedia*.

[2] *New Technologies in Protective Coatings, Guadalupe Canosa, Ed Carlos GiudiceIntechOpen, 2017*

[3] Weil, E. D.; Levchik, S. In *Flame Retardants for Plastics and Textiles: Practical Applications*, 2009.

[4] McKenna, S. T.; Birtles, R.; Dickens, K.; Walker, R. G.; Spearpoint, M. J.; Stec, A. A.; Hull, T. R., Flame retardants in UK furniture increase smoke toxicity more than they reduce fire growth rate. *Chemosphere* **2018,** *196*, 429-439.

[5] Brown, P.; Cordner, A., Lessons learned from flame retardant use and regulation could enhance future control of potentially hazardous chemicals. *Health affairs (Project Hope)* **2011,** *30* (5), 906-14.

[6] G.J. van Esch, B., United Nations Environment programme international labour organization world health organaization *International programme on chemical saftey* **1997**.

[7] Kausar, A.; Rafique, I.; Anwar, Z.; Muhammad, B., Recent Developments in Different Types of Flame Retardants and Effect on Fire Retardancy of Epoxy Composite. *Polymer-Plastics Technology and Engineering* **2016,** *55* (14), 1512-1535.

[8] Sinha Ray S., Kuruma M. (2020) Types of Flame Retardants Used for the Synthesis of Flame-Retardant Polymers. In: Halogen-Free Flame-Retardant Polymers. Springer Series in Materials Science, vol 294. Springer, Cham. https://doi.org/10.1007/978-3-030-35491-6_4.

[9] Qiu, Y.; Qian, L.; Feng, H.; Jin, S.; Hao, J., Toughening Effect and Flame-Retardant Behaviors of Phosphaphenanthrene/Phenylsiloxane Bigroup Macromolecules in Epoxy Thermoset. *Macromolecules* **2018,** *51* (23), 9992-10002.

[10] Tawfik, S. Y., Flame Retardants: Additives in Plastic Technology. In *Polymers and Polymeric Composites: A Reference Series*, Palsule, S., Ed. Springer Berlin Heidelberg: Berlin, Heidelberg, 2017; pp 1-27.

[11] Lu, S.; Hong, W.; Chen, X., Nanoreinforcements of Two-Dimensional Nanomaterials for Flame Retardant Polymeric Composites: An Overview. *Advances in Polymer Technology* **2019,** *2019*, 4273253.

[12] V.G. Schevchuk; E.N. Kondratev; A.N. Zolotko; Smirnov, V. V., Conditions of flame propagation in aerosuspensions of metallic particles *Comb. Expl. Shock Waves,* **1983,** *82*, 557-561.

[13] Raman, V.; Fox, R. O., Modeling of Fine-Particle Formation in Turbulent Flames. *Annual Review of Fluid Mechanics* **2016,** *48* (1), 159-190.

[14] Suriyawong, A.; Gamble, M.; Lee, M.-H.; Axelbaum, R.; Biswas, P., Submicrometer Particle Formation and Mercury Speciation Under O2–CO2 Coal Combustion. *Energy & Fuels* **2006,** *20* (6), 2357-2363.

[15] Sung, Y.; Raman, V.; Fox, R. O., Large-eddy-simulation-based multiscale modeling of TiO2 nanoparticle synthesis in a turbulent flame reactor using

detailed nucleation chemistry. *Chemical Engineering Science* **2011,** *66* (19), 4370-4381.

[16] Marc D. Rumminger; Linteris, G. T., The role of particles in the inhibition of pre-mixed flames by iron pentacarbonyl. *Combustion and Flame* **2000,** *123* (1), 82-94.

[17] Hoffman, K.; Sosa, J. A.; Stapleton, H. M., Do flame retardant chemicals increase the risk for thyroid dysregulation and cancer? *Current opinion in oncology* **2017,** *29* (1), 7-13.

[18] Molyneux, S.; Stec, A. A.; Hull, T. R., The effect of gas phase flame retardants on fire effluent toxicity. *Polymer Degradation and Stability* **2014,** *106*, 36-46.

[19] Fent, K. W.; LaGuardia, M.; Luellen, D.; McCormick, S.; Mayer, A.; Chen, I. C.; Kerber, S.; Smith, D.; Horn, G. P., Flame retardants, dioxins, and furans in air and on firefighters' protective ensembles during controlled residential firefighting. *Environment International* **2020,** *140*, 105756.

[20] Rodgers, K. M.; Covaci, A.; Poma, G.; Knox, K.; Allen, J. G.; Cedeno-Laurent, J.; Rudel, R. A.; Dodson, R. E., Flame Retardant Concentrations Are Lower in College Spaces Meeting the New Furniture Flammability Standard TB117-2013. *Environmental Science & Technology Letters* **2020.**

[21] Dodson, R. E.; Rodgers, K. M.; Carey, G.; Cedeno Laurent, J. G.; Covaci, A.; Poma, G.; Malarvannan, G.; Spengler, J. D.; Rudel, R. A.; Allen, J. G., Flame Retardant Chemicals in College Dormitories: Flammability Standards Influence Dust Concentrations. *Environmental Science & Technology* **2017,** *51* (9), 4860-4869.

[22] Waaijers, S. L.; Kong, D.; Hendriks, H. S.; de Wit, C. A.; Cousins, I. T.; Westerink, R. H.; Leonards, P. E.; Kraak, M. H.; Admiraal, W.; de Voogt, P.; Parsons, J. R., Persistence, bioaccumulation, and toxicity of halogen-free flame retardants. *Reviews of environmental contamination and toxicology* **2013,** *222*, 1-71.

[23] Juleen Lam; Bruce P. Lanphear; David Bellinger; Daniel A. Axelrad; Jennifer McPartland; Patrice Sutton; Lisette Davidson; Natalyn Daniels; Saunak Sen; Woodruff, T. J., Developmental PBDE Exposure and IQ/ADHD in Childhood: A Systematic Review and Meta-analysis. *Environmental Health Perspectives* **2017,** *125* (8), 1.

[24] Canbaz, D.; Logiantara, A.; van Ree, R.; van Rijt, L. S., Immunotoxicity of organophosphate flame retardants TPHP and TDCIPP on murine dendritic cells in vitro. *Chemosphere* **2017,** *177*, 56-64.

[25] (a) Greeson, K. W.; Fowler, K. L.; Estave, P. M.; Kate Thompson, S.; Wagner, C.; Clayton Edenfield, R.; Symosko, K. M.; Steves, A. N.; Marder, E. M.; Terrell, M. L.; Barton, H.; Koval, M.; Marcus, M.; Easley, C. A., Detrimental effects of flame retardant, PBB153, exposure on sperm and future generations. *Scientific Reports* **2020,** *10* (1), 8567;(b) Betts, K. S., Endocrine damper? Flame retardants linked to male hormone, sperm count changes. *Environ Health Perspect* **2010,** *118* (3), A 130.

[26] Tian, J.; Ji, Z.; Wang, F.; Song, M.; Li, H., The Toxic Effects of Tetrachlorobisphenol A in *Saccharomyces cerevisiae* Cells via Metabolic Interference. *Sci Rep* **2017,** *7* (1), 2655.

[27] Araki, A.; Saito, I.; Kanazawa, A.; Morimoto, K.; Nakayama, K.; Shibata, E.; Tanaka, M.; Takigawa, T.; Yoshimura, T.; Chikara, H.; Saijo, Y.; Kishi, R., Phosphorus flame retardants in indoor dust and their relation to asthma and allergies of inhabitants. *Indoor Air* **2014,** *24* (1), 3-15.

[28] Guigueno, M. F.; Fernie, K. J., Birds and flame retardants: A review of the toxic effects on birds of historical and novel flame retardants. *Environmental research* **2017,** *154*, 398-424.

[29] Khaled, A.; Rivaton, A.; Richard, C.; Jaber, F.; Sleiman, M., Phototransformation of Plastic Containing Brominated Flame Retardants: Enhanced Fragmentation and Release of Photoproducts to Water and Air. *Environmental Science & Technology* **2018,** *52* (19), 11123-11131.

[30] (a) Castro-Jiménez, J.; González-Gaya, B.; Pizarro, M.; Casal, P.; Pizarro-Álvarez, C.; Dachs, J., Organophosphate Ester Flame Retardants and Plasticizers in the Global Oceanic Atmosphere. *Environmental Science & Technology* **2016,** *50* (23), 12831-12839;(b) Zhang, Q.; Lu, M.; Dong, X.; Wang, C.; Zhang, C.; Liu, W.; Zhao, M., Potential Estrogenic Effects of Phosphorus-Containing Flame Retardants. *Environmental Science & Technology* **2014,** *48* (12), 6995-7001;(c) Liu, L.-Y.; He, K.; Hites, R. A.; Salamova, A., Hair and Nails as Noninvasive Biomarkers of Human Exposure to Brominated and Organophosphate Flame Retardants. *Environmental Science & Technology* **2016,** *50* (6), 3065-3073;(d) Zhang, Q.; Yu, C.; Fu, L.; Gu, S.; Wang, C., New Insights in the Endocrine Disrupting Effects of Three Primary Metabolites of Organophosphate Flame Retardants. *Environmental Science & Technology* **2020,** *54* (7), 4465-4474.

[31] Babrauskas, V.; Fuoco, R.; Blum, A., Chapter 3 - Flame Retardant Additives in Polymers: When do the Fire Safety Benefits Outweigh the Toxicity Risks? In *Polymer Green Flame Retardants*, Papaspyrides, C. D.; Kiliaris, P., Eds. Elsevier: Amsterdam, 2014; pp 87-118.

[32] Innes, A.; Innes, J., 10 - Flame Retardants. In *Handbook of Environmental Degradation of Materials (Second Edition)*, Kutz, M., Ed. William Andrew Publishing: Oxford, 2012; pp 309-335.

[33] Hull, T. R.; Law, R. J.; Bergman, Å., Chapter 4 - Environmental Drivers for Replacement of Halogenated Flame Retardants. In *Polymer Green Flame Retardants*, Papaspyrides, C. D.; Kiliaris, P., Eds. Elsevier: Amsterdam, 2014; pp 119-179.

[34] Morgan, A. B., The Future of Flame Retardant Polymers – Unmet Needs and Likely New Approaches. *Polymer Reviews* **2019,** *59* (1), 25-54.

[35] Polymers., D.-P. A., Nanoparticle Reinforced Polymers. *Polymers* **2019,** *11* (4), 625.

[36] Wilson, W. E.; Fristrom, R. M., Radicals in Flames. *APL Technical Digest* **1963,** 7.

[37] Markwart, J. C.; Battig, A.; Zimmermann, L.; Wagner, M.; Fischer, J.; Schartel, B.; Wurm, F. R., Systematically Controlled Decomposition Mechanism in Phosphorus Flame Retardants by Precise Molecular Architecture: P–O vs P–N. *ACS Applied Polymer Materials* **2019,** *1* (5), 1118-1128.

[38] Pielichowski, K.; Njuguna, J., *Thermal Degradation Of Polymeric Materials*. 2008.

[39] H., N., Polymers in everyday life. *BioImpacts* **2017,** *7* (2), 73-74.

[40] Zhang, H., " (2003). Doctoral Dissertations 1896 "Fire-safe polymers and polymer composites. **2003,** 1053.

[41] Álvarez-Muñoz, D.; Llorca, M.; Blasco, J.; Barceló, D., Chapter 1 - Contaminants in the Marine Environment. In *Marine Ecotoxicology*, Blasco, J.; Chapman, P. M.; Campana, O.; Hampel, M., Eds. Academic Press: 2016; pp 1-34.

[42] Zheng, L., 1 - Overview of oxy-fuel combustion technology for carbon dioxide (CO2) capture. In *Oxy-Fuel Combustion for Power Generation and Carbon Dioxide (CO2) Capture*, Zheng, L., Ed. Woodhead Publishing: 2011; pp 1-13.

[43] (a) Koncar, V., 1 - Smart textiles for monitoring and measurement applications. In *Smart Textiles for In Situ Monitoring of Composites*, Koncar, V., Ed. Woodhead Publishing: 2019; pp 1-151;(b) Kumar, A.; T'ien, J. S., Numerical Modeling of Limiting Oxygen Index Apparatus for Film Type Fuels. *International Journal of Spray and Combustion Dynamics* **2012,** *4* (4), 299-322.

[44] Liu, C.; Fu, L.; Yang, J.; Zhang, S.; Shi, Y.; Yang, F.; Yang, Y.; Zhuang, Y.; Ye, Y.; Wang, L., A novel understanding of combustion behavior of coals by cone calorimeter. *Journal of Thermal Analysis and Calorimetry* **2020**.

[45] Ellzey, K. A.; Ranganathan, T.; Zilberman, J.; Coughlin, E. B.; Farris, R. J.; Emrick, T., Deoxybenzoin-Based Polyarylates as Halogen-Free Fire-Resistant Polymers. *Macromolecules* **2006,** *39* (10), 3553-3558.

[46] Solomon, P. R.; Hamblen, D. G.; Carangelo, R. M.; Serio, M. A.; Deshpande, G. V., General model of coal devolatilization. *Energy & Fuels* **1988,** *2* (4), 405-422.

[47] Hambly, E. M., *Thesis entitled "The Chemical Structure of Coal Tar and Char During Devolatilization" (1998)*.

[48] Crosby, A. J.; Lee, J. Y., Polymer Nanocomposites: The "Nano" Effect on Mechanical Properties. *Polymer Reviews* **2007,** *47* (2), 217-229.

[49] Kashiwagi, T.; Du, F.; Douglas, J. F.; Winey, K. I.; Harris, R. H.; Shields, J. R., Nanoparticle networks reduce the flammability of polymer nanocomposites. *Nature Materials* **2005,** *4* (12), 928-933.

[50] Burgos-Mármol, J. J.; Álvarez-Machancoses, Ó.; Patti, A., Modeling the Effect of Polymer Chain Stiffness on the Behavior of Polymer Nanocomposites. *The Journal of Physical Chemistry B* **2017,** *121* (25), 6245-6256.

[51] Aseeva, R. M.; Zaikov, G. E. In *Flammability of polymeric materials*, Berlin, Heidelberg, Springer Berlin Heidelberg: Berlin, Heidelberg, 1985; pp 171-229.

[52] Marlair, G.; Tewarson, A., . Effects Of The Generic Nature Of Polymers On Their Fire Behavior. *Fire Safety Science* **2003,** *7*, 629-642.

[53] Petrella, R. V., Factors Affecting the Combustion of Polystyrene and Styrene. In *Flame - Retardant Polymeric Materials: Volume 2*, Lewin, M.; Atlas, S. M.; Pearce, E. M., Eds. Springer US: Boston, MA, 1978; pp 159-201.

[54] Malkappa, K.; Ray, S. S.; Kumar, N., Enhanced Thermo-Mechanical Stiffness, Thermal Stability, and Fire Retardant Performance of Surface-Modified 2D MoS2 Nanosheet-Reinforced Polyurethane Composites. *Macromolecular Materials and Engineering* **2019,** *304* (1), 1800562.

[55] Varol, H. S.; Meng, F.; Hosseinkhani, B.; Malm, C.; Bonn, D.; Bonn, M.; Zaccone, A.; Parekh, S. H., Nanoparticle amount, and not size, determines chain alignment and nonlinear hardening in polymer nanocomposites. *Proceedings of the National Academy of Sciences* **2017,** *114* (16), E3170-E3177.

[56] Oguz, O.; Candau, N.; Citak, M. K.; Cetin, F. N.; Avaz Seven, S.; Menceloglu, Y. Z., A Sustainable Approach to Produce Stiff, Super-Tough, and Heat-Resistant Poly(lactic acid)-Based Green Materials. *ACS Sustainable*

Chemistry & Engineering **2019,** *7* (8), 7869-7877.

[57] Bhattacharya, M., Polymer Nanocomposites-A Comparison between Carbon Nanotubes, Graphene, and Clay as Nanofillers. *Materials* **2016,** (9), 262.

[58] Yook, S.; Shams Es-haghi, S.; Yildirim, A.; Mutlu, Z.; Cakmak, M., Anisotropic hydrogels formed by magnetically-oriented nanoclay suspensions for wound dressings. *Soft Matter* **2019,** *15* (47), 9733-9741.

[59] T. J. Pinnavaia and G. W. Beall are the authors of Polymer-Clay Nanocomposites, published by Wiley, 2000.

[60] Cadambi, R. M.; Ghassemieh, E., Mechanism of Gas Permeation in Processed HNBR/Nanoclay Composites. *Advances in Polymer Technology* **2013,** *32* (S1), E103-E118.

[61] Maria, H. J.; Thomas, M. G.; Morreale, M.; La Mantia, F. P.; Nzihou, A.; Joseph, K.; Rouxel, D.; Fernandes, S. C. M.; Kalarikkal, N.; Thomas, S., Gas Barrier, Rheological and Mechanical Properties of Immiscible Natural Rubber/Acrylonitrile Butadiene Rubber/Organoclay (NR/NBR/Organoclay) Blend Nanocomposites. *Materials (Basel)* **2020,** *13* (11), 2654.

[62] Ben Dhieb, F.; Dil, E. J.; Tabatabaei, S. H.; Mighri, F.; Ajji, A., Effect of nanoclay orientation on oxygen barrier properties of LbL nanocomposite coated films. *RSC Advances* **2019,** *9* (3), 1632-1641.

[63] Das, P.; Malho, J.-M.; Rahimi, K.; Schacher, F. H.; Wang, B.; Demco, D. E.; Walther, A., Nacre-mimetics with synthetic nanoclays up to ultrahigh aspect ratios. *Nature Communications* **2015,** *6* (1), 5967.

[64] Lu, C.; Mai, Y.-W., Influence of Aspect Ratio on Barrier Properties of Polymer-Clay Nanocomposites. *Physical Review Letters* **2005,** *95* (8), 088303.

[65] Horrocks, A. R., Developments in flame retardants for heat and fire resistant textiles—the role of char formation and intumescence. *Polymer Degradation and Stability* **1996,** *54* (2), 143-154.

[66] Wang, Z.-y.; Han, E.-h.; Ke, W., Fire-resistant effect of nanoclay on intumescent nanocomposite coatings. *Journal of Applied Polymer Science* **2007,** *103* (3), 1681-1689.

[67] Salih, W. K., Fire retardancy assessment of polypropylene composite filed with nano clay prepared from Iraqi bentonite. *Journal of Physics: Conference Series* **2018,** *1003*, 012019.

[68] Xu, X.; Ding, Y.; Qian, Z.; Wang, F.; Wen, B.; Zhou, H.; Zhang, S.; Yang, M., Degradation of poly(ethylene terephthalate)/clay nanocomposites during melt extrusion: Effect of clay catalysis and chain extension. *Polymer Degradation and Stability* **2009,** *94* (1), 113-123.

[69] Flame Retardant Mechanism of Polymer–Clay Nanocomposites. In *Flame Retardant Polymer Nanocomposites*, 2007; pp 67-87.

[70] Huang, G.; Yang, J.; Wang, X.; Gao, J., Nanoclay, intumescent flame retardants, and their combination with chemical modification for the improvement of the flame retardant properties of polymer nanocomposites. *Macromolecular Research* **2013,** *21* (1), 27-34.

[71] Zhang, J.; Wu, Q.; Li, G.; Li, M.-C.; Sun, X.; Ring, D., Synergistic influence of halogenated flame retardants and nanoclay on flame performance of high density polyethylene and wood flour composites. *RSC Advances* **2017,** *7* (40), 24895-24902.

[72] Bellucci, F.; Camino, G.; Frache, A.; Sarra, A., Catalytic charring–volatilization competition in organoclay nanocomposites. *Polymer Degradation and Stability* **2007,** *92* (3), 425-436.

[73] Gilman, J. W.; Jackson, C. L.; Morgan, A. B.; Harris, R.; Manias, E.; Giannelis, E. P.; Wuthenow, M.; Hilton, D.; Phillips, S. H., Flammability Properties of Polymer−Layered-Silicate Nanocomposites. Polypropylene and Polystyrene Nanocomposites. *Chemistry of Materials* **2000,** *12* (7), 1866-1873.

[74] Eckert, A.; Abbasi, M.; Mang, T.; Saalwächter, K.; Walther, A., Structure, Mechanical Properties, and Dynamics of Polyethylenoxide/Nanoclay Nacre-Mimetic Nanocomposites. *Macromolecules* **2020,** *53* (5), 1716-1725.

[75] Giannelis, E. P., Polymer-layred silicate nanocomposites: Synthesis, properties and applications. *Appl. Organometal. Chem.* **1998,** *12*, 675-688.

[76] Khatoon, N.; Chu, M. Q.; Zhou, C. H., Nanoclay-based drug delivery systems and their therapeutic potentials. *Journal of Materials Chemistry B* **2020,** *8* (33), 7335-7351.

[77] Lv, W.-F.; Zhou, Z.-H.; Zhang, Q.; Luo, W.-L.; Wang, H.-Z.; Ma, D.-S.; Zhang, L.; Wang, R.; Zhang, L., Wetting of polymer surfaces by aqueous solutions of branched cationic Gemini surfactants. *Soft Matter* **2019,** *15* (33), 6725-6731.

[78] Schartel, B.; Knoll, U.; Hartwig, A.; Pütz, D., Phosphonium-modified layered silicate epoxy resins nanocomposites and their combinations with ATH and organo-phosphorus fire retardants. *Polymers for Advanced Technologies* **2006,** *17* (4), 281-293.

[79] Jong-Ho Back; Dooyoung Baek; Taeyoon Kim; Bongkuk Seo; Wonjoo Lee; Kim, H.-J., Synthesis of phosphorus-containing polyol and its effects on impact resistance and flame retardancy of structural epoxy adhesives. *International Journal of Adhesion & Adhesives* **2020,** *100*, 102601.

[80] Wang, X.; Zhang, P.; Huang, Z.; Xing, W.; Song, L.; Hu, Y., Effect of aluminum diethylphosphinate on the thermal stability and flame retardancy of flexible polyurethane foams. *Fire Safety Journal* **2019,** *106*, 72-79.

[81] Braun, U.; Schartel, B.; Fichera, M. A.; Jäger, C., Flame retardancy mechanisms of aluminium phosphinate in combination with melamine polyphosphate and zinc borate in glass-fibre reinforced polyamide 6,6. *Polymer Degradation and Stability* **2007,** *92* (8), 1528-1545.

[82] Kaynak, C.; Polat, O., Influences of nanoclays on the flame retardancy of fiber-filled and unfilled polyamide-6 with and without aluminum diethylphosphinate. *Journal of Fire Sciences* **2015,** *33* (2), 87-112.

[83] Chen, J.; Han, J., Comparative performance of carbon nanotubes and nanoclays as flame retardants for epoxy composites. *Results in Physics* **2019,** *14*, 102481.

[84] Choi, D.; Sonkaria, S.; Fox, S. J.; Poudel, S.; Kim, S.-y.; Kang, S.; Kim, S.; Verma, C.; Ahn, S. H.; Lee, C. S.; Khare, V., Quantum scale biomimicry of low dimensional growth: An unusual complex amorphous precursor route to TiO2 band confinement by shape adaptive biopolymer-like flexibility for energy applications. *Scientific Reports* **2019,** *9* (1), 18721.

[85] Deka, B. K.; Maji, T. K., Effect of TiO2 and nanoclay on the properties of wood polymer nanocomposite. *Composites Part A: Applied Science and Manufacturing* **2011,** *42* (12), 2117-2125.

[86] Suter, J. L.; Groen, D.; Coveney, P. V., Mechanism of Exfoliation and

Prediction of Materials Properties of Clay–Polymer Nanocomposites from Multiscale Modeling. *Nano Letters* **2015,** *15* (12), 8108-8113.

[87] Lee, J.-H.; Park, J.-W.; Kim, H.; Jang, S.-W.; Kim, H.-J.; Choi, Y., Thermal property and flame retardancy comparisons based on particle size and size distribution of clays in ethylene vinyl acetate (EVA) adhesive sheets for cross-laminated timber (CLT). *European Journal of Wood and Wood Products* **2020,** *78* (1), 93-105.

[88] Kim, H.; Park, J.-W.; Lee, J.-H.; Jang, S.-W.; Kim, H.-J.; Choi, Y.; Choy, J.-H.; Yang, J.-H., Clay-organic intumescent hybrid system for the synergetic flammability of polymer nanocomposites. *Journal of Thermal Analysis and Calorimetry* **2018,** *132* (3), 2009-2014.

[89] Qin, H.; Zhang, S.; Zhao, C.; Hu, G.; Yang, M., Flame retardant mechanism of polymer/clay nanocomposites based on polypropylene. *Polymer* **2005,** *46* (19), 8386-8395.

[90] Rapacz-Kmita, A.; Gajek, M.; Dudek, M.; Stodolak-Zych, E.; Szaraniec, B.; Lach, R., Thermal, structural and mechanical analysis of polymer/clay nanocomposites with controlled degradation. *Journal of Thermal Analysis and Calorimetry* **2017,** *127* (1), 389-398.

[91] Liu, W.; Varley, R. J.; Simon, G. P., Understanding the decomposition and fire performance processes in phosphorus and nanomodified high performance epoxy resins and composites. *Polymer* **2007,** *48* (8), 2345-2354.

[92] Saha, T.; Bhowmick, A. K.; Oda, T.; Miyauchi, T.; Fujii, N., Influence of layered nanofillers on the mechanical properties and thermal degradation of polyacrylicester polymer: Theoretical and experimental investigations. *Composites Part B: Engineering* **2019,** *169*, 65-78.

[93] Keshavarz, M. H., *Combustible Organic Materials: Determination and Prediction of Combustion Properties*. De Gruyter: 2018.

[94] Xu, Y.; Li, S.; Gao, Q.; Yao, Q.; Liu, J., Characterization on Ignition and Volatile Combustion of Dispersed Coal Particle Streams: In Situ Diagnostics and Transient Modeling. *Energy & Fuels* **2018,** *32* (9), 9850-9858.

[95] Li, J.; Paul, M. C.; Younger, P. L.; Watson, I.; Hossain, M.; Welch, S., Prediction of high-temperature rapid combustion behaviour of woody biomass particles. *Fuel* **2016,** *165*, 205-214.

[96] Xu, Q.; Jin, C.; Jiang, Y., Compare the flammability of two extruded polystyrene foams with micro-scale combustion calorimeter and cone calorimeter tests. *Journal of Thermal Analysis and Calorimetry* **2017,** *127* (3), 2359-2366.

Chapter 3

Development of Halogen Free Sustainable Polybenzoxazine Matrices and Composites for Flame Retardant Applications

Subramani Devaraju, Krishnamoorthy Krishnadevi and Muthukaruppan Alagar

Abstract

In recent years, with the growing concern on energy crises, the development of new products from sustainable resources have been received much greater consideration owing to the environmental concerns caused by the rapid reduction of non-renewable fossil resources. The widespread utilization of polymeric materials led to a considerable increase in fire risk in our daily life due to their flammable behavior. Hence, the flame-retardant properties of polymeric materials are considered as one of the most significant criteria to use them in wide range of industrial applications. It is well known that mostly halogenated flame retardant materials are widely used for flame retardant applications. However, they release toxic gases and corrosive smoke during combustion, and consequently they contribute to serious environmental pollution. In this connection, in the present chapter we discussed the different types of non-halogen based environmentally friendly bio-based polybenzoxazine matrices developed from renewable and sustainable bio-phenolic materials viz., cardanol, eugenol, guiacol and etc., and the composites obtained by reinforcing with varying weight percentages of flame resistant reinforcements/additives and their flame retardant properties obtained from different analytical techniques.

Keywords: Bio-phenols, sustainable polybenzoxazine, bio-based composites, flame retardancy, halogen free benzoxazine composites

1. Introduction

Most of the polymeric materials including synthetic and natural based, are extremely flammable in nature. The extensive usage of polymer based materials led to a substantial increase in fire risk in our day to day-life. For example, about 2 million fire accidents are reported in Europe every year, which cause more than about 4000 casualties and ~ 70000 injuries along with economic loss of about ~120 billion € [1, 2]. Thus, there are unwanted and excessive societal and economic burdens on the entire civilization, hence it is most essential to produce polymeric products with efficient flame retardant properties to avoid fire related losses.

In addition, the flame-retardant behaviour of polymeric materials are considered as one of the most important criteria to utilize them in the form of sealants, encapsulants, coatings, and matrices for different industrial applications [3–8].

In the recent years, due to the environmental, sustainable and economic aspect, growing attempts have been progressed for the synthesis of polymeric resin from natural renewable/sustainable feedstock instead of using fossil fuel/petroleum based raw materials [9–12]. The production of polymeric materials at present are mostly based on the petroleum feedstock and are used for wide range of industrial applications [13–16]. Bio-based matrices/composites already found numerous applications in the diverse fields of our everyday lives, such as in the automotive industry or in building and construction, aerospace, and so forth [17–23]. In this context, in order to improve their safe utility and also widen their area of application, it is essential to increase their fire/flame resistant behavior to make them suitable for indented applications. Hence, the development of polymeric materials with enhanced flame retardant behavior is warranted from sustainable bio-resources.

Usually, halogen based flame resistant additive materials has been extensively used into polymer matrix to enhance its flame resistant properties. The release the corrosive smoke and toxic gases by these additives during pyrolysis, consequently, give rise to serious ecological problems. Consequently, the development of polymeric materials free from halogens are considered to be a versatile approach. The method of enhancing the flame resistant properties of polymers without use of halogen based flame additives can be deliberated by the following three methodologies to obtain the flame retardant of polymers: (i) using intrinsically flame retardant polymeric materials, (ii) to suitable structural modification of existing industrial polymers, (iii) to introduce intrinsic flame resistant nano-reinforcements/fillers into polymers.

As a result, new research fields have been emerged in the material science in particular in the field of polymers and directed toward a completely or partial replacement of the petroleum-based materials. In this regard, polybenzoxazines possess very useful perspective due to their extraordinary molecular design flexibility [24–32], which easily facilitates and allows bio-based precursors for the production of varied molecular structured benzoxazines [33–57], consequently reducing/replacing the considerable utilization of the petroleum-based raw materials. The nitrogen atom in the skeleton of benzoxazines significantly enhances their flame retardancy, making them suitable for the fabrication of flame retardant components. Fire, Smoke, and Toxicity (FST) reduction behaviour of benzoxazine based materials have attracted industrial and academic attention as federal regulations become strict and new technologies emerge. It was reported that the many efforts have been made to syntheses of benzoxazines exploiting natural renewable resources using cardanol, urushiol, coumarine, eugenol, guaiacol, furfuryl amine and stearyl amine. In addition, silica/phosphorous/carbon reinforced benzoxazine hybrid materials expected to possess radiation resistant behaviour coupled with inherent flame retardant properties. Probable flame resistant mechanism of the polybenzoxazine resin is owing to the release of gaseous species on the surface. The gases including CO_2, NO_2, and H_2O might able to diminish the O_2 concentration around the burning area and diminish the heat of the surface. Another probable mechanism of PBz polymer, with flame resistant layer through the formation of char on the surface. The inert flame resistant layer might protect the benzoxazine from the external heat from the flame, thus performing as the O_2 protection layer. The high strength, excellent flame retardancy, radiation resistance, high thermal stability, low moisture absorption, low temperature cure, low shrinkage and low-k dielectric behaviour have made benzoxazine resins become an attractive for electronics and aerospace applications. Hence, in the present chapter, the preparation, properties of some of the important and valuable halogen free bio-based flame retardant polybenzoxazines matrices and composites are reviewed and summarized.

2. Cardanol based polybenzoxazines for flame resistant applications

Cardanol is a naturally occurring combination of bio-phenolic materials isolated from cashew nut-shell agro- waste with exported worth of 1.39 USD million in the year 2020–2021 (Apr-Nov) by India [58]. The cardanol chemical structure is very exciting due to its reactive functional phenolic-OH group and an alkyl spacer with unsaturation in m-position will be applicable for several chemical reaction and functional group modifications [59–62]. Cardanol bargains one of the better probable material for green synthesis of benzoxazines due to their massive availability and very low cost of production. Also, cardanol competently alternative to petro-phenol (for example bisphenol-A (BPA) which is an endocrine disruptor) to prepare BPA-free polybenzoxazines. There are numerous auspicious ingenuities for replacing the petro-based phenolic resources with cardanol to form cardanol based polybenzoxazines.

The flame retardant properties of cardanol based polybenzoxazines are usually lower than those of petroleum based thermosets. However, solutions can be provided by either physical blending with flame retardants or by chemical modification. Several research groups attempted to solve these issues by designing multi amine precursors [63, 64] or by introducing, silica/silicon, phosphorus, boron, and furan [52, 65–70].

To improve the flame resistant properties, Bimlesh Lochab research group introduced the [71] halogen free eco-friendly hexa-functional cardanol (bio-based phenolic) benzoxazine with a phosphazene core (cyclophosphazene ring based benzoxazine monomer designated as CPN) as reactive flame retardant precursor. The CPN monomer showed good compatibility with cardanol and tris-p-aminophenylmethane based benzoxazine monomer (CPN_0). Higher loading of CPN in the monomer improved both the smoke density rating, vertical burning rating and also led to higher limiting oxygen index (LOI). Finally, the introduction of CPN network shows good compatibility with the polybenzoxazine phenolic thermosets with enhancement in flame resistant properties.

CPN monomer was synthesized as per the reported procedure (**Figure 1**). The phosphazene core with hexa-amine (2) was synthesised using phosphonitrilic chloride trimer ($N_3P_3Cl_6$), by base facilitated reaction with excess of 4-acetomidophenol to yield compound (1) and, which upon hydrolysis, gave the corresponding hexa-amine compound (2). Later, the synthesized hexa-amine (2) was reacted with 4-pentadecylsalicylaldehyde (3) to form the corresponding Schiff base which is then reduced to compound (4). Compound (4) undergoes intra-molecular cyclization reaction to form CPN (5). The renewable cardanol and phosphorous content in the CPN monomer are 65.7% and 3.4% respectively.

The cardanol (CPN_0) monomer was synthesised as per the procedure reported [71]. The monomer combinations are named as CPNx, here x is percentage weight of CPN incorporated in the CPN_0 monomer and blends developed are named as CPN_{10}, and CPN_{80}. The polymerization of CPN, CPN_{80}, CPN_{10}, andCPN_0 was accomplished in a hot air oven at the temperature of 50°C, 100°C, 120°C, 150°C, 180°C, 200°C, 220°C, 240°C for each 1 h and followed by further heating at 240 °C for 0 h, 1 h, 1.5 h and 2 h. The prepared poly (CPN_X) (**Figure 2**) was utilized for further characterizations.

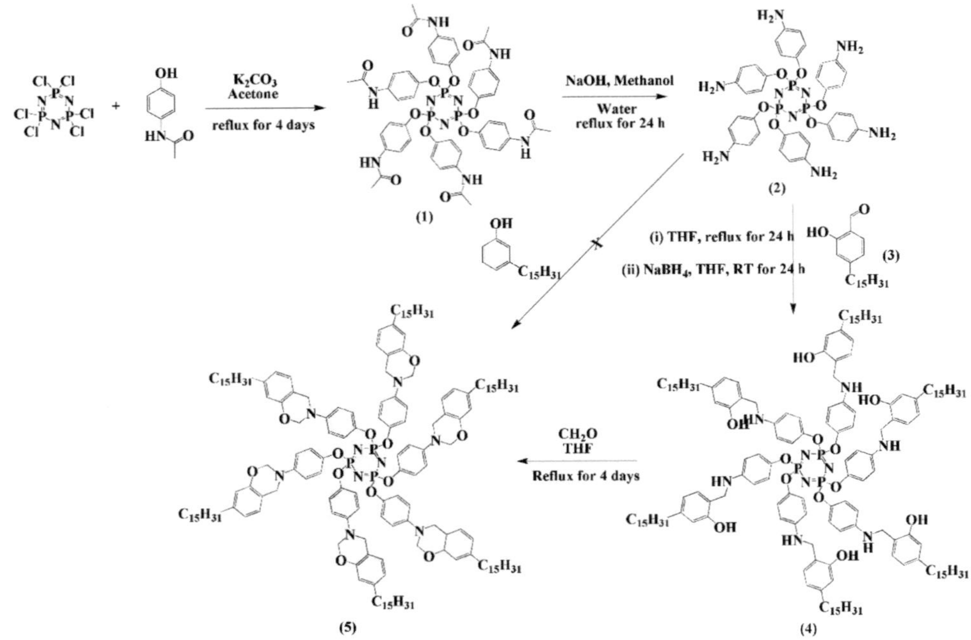

Figure 1.
Divergent approach for the synthesis of cardanol phosphazene benzoxazine monomer (CPN).

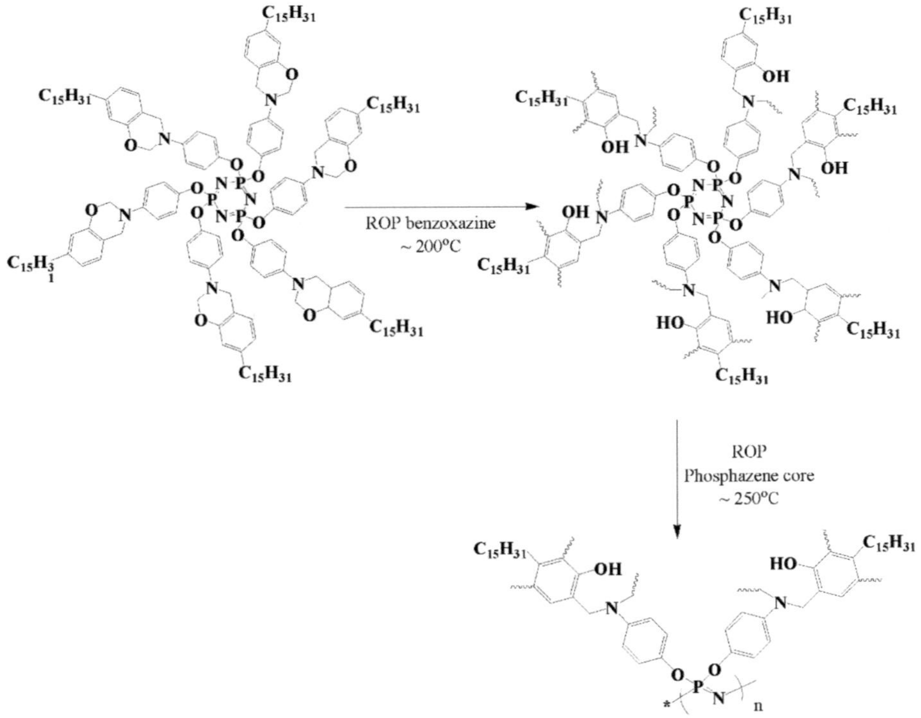

Figure 2.
Ring opening polymerisation of CPN monomer to form polyphosphazene polybenzoxazine.

The limiting oxygen index (LOI) of the polymers was calculated from TGA using van Krevelen's equation and the values obtained are presented in **Table 1**. If a polymer possesses the value of LOI is less than 20.9%, it burn simply in air; 21–28%

Samples	Phosphorus (%)	Cardanol (%)	Char(%) residue	LOI	UL-94	Smoke density	Reference
CPN_0	0	70.40	14.0 (0^a)	23 (18^a)	V-2	77.70	[71]
CPN_{10}	0.30	70.00	24 .0(7^a)	27 (20^a)	V-1	70.70	[71]
CPN_{80}	2.70	66.80	32.0(26^a)	30 (28^a)	V-0	34.30	[71]
CPN	3.40	65.70	39.0(28^a)	33 (29^a)	V-0	33.90	[71]
C-trisapm (T)	0	70.40^b	8.30	20.82		54.60	[72]
CP_1T_3	1.19	23.28^b	28.50	28.90		51.60	[72]
CP_3T_1	3.56	69.83^b	42.70	34.58		47.50	[72]
CP	4.79	93.10^b	29.30	29.22		34.01	[72]
EP_1T_3	2.09	21.97^b	28.70	28.98		35.39	[72]
EP_3T_1	6.26	65.92^b	28.50	28.9		39.83	[72]
EP	8.34	87.89^b	32.40	30.46		18.80	[72]

[a]*Air atmosphere.*
[b]*Renewable phenol (%).*

Table 1.
Thermal and flame resistant (LOI, UL-94 and smoke density) properties of bio-based polybenzoxazine matrices and composites.

slowburning; 28 to 100% self-extinguishing; and > 100 is considered as inherently non-flammable [71]. From the **Table 1**, an increment in the value of LOI is reliant on the weight percentage content of phosphorous signifying the role played by CPN core in flammability. The vertical burning test (UL-94) was adopted to check the flammability features of poly(CPNx) and the values obtained are presented in **Table 1**. The poly(CPN_0) was burnt immediately with a lower combustion time along with fire drippings. The flame resistance properties was increased with an increase in weight content of phosphorus which substantiates with LOI results.

A smoke density test was carried out to ascertain the relative quantities of smoke produced during burning of poly(CPN_x). The neat poly(CPN_0) indicated nearly very low residue char which crumbled in the mesh, whereas phosphazene containing poly(CPN_{10}), poly(CPN_{80}), and poly(CPN) exhibited high char residue and perceived smoke density rating of 77.7, 70.8, 34.3 and 33.9 respectively. The comparative variance in smoke density was observed as 6.9 and 43.4 with 10 wt% and 80 wt.% respectively with introduction of CPN proposing a substantial role frolicked by CPN core in the decrease of smoke density in case of poly(CPN_0).

Further to confirm the flame retardant mechanism, SEM was performed to check the morphology of the residual char samples after smoke density analysis and are depicted in **Figure 3**. From SEM images, it was noticed that the several undulations and ripples in the cases of poly(CPN_0), and poly(CPN_{10}) due to the presence of a higher amount of flammable matrix (**Figure 3a** and **b**). The formation of bigger cracks (20–30 μm) on exterior surface was observed in the residue of poly(CPN_0). The cracks were significantly decreased with the introduction of CPN, as observed from the exterior surfaces of poly(CPN_{10}), poly(CPN_{80}), and poly(CPN). Both poly (CPN_{80}), and poly(CPN) formed a extremely compact and thick charred layers with combined enlarged residual char. Analysis of interior surface morphology exhibited the development of bigger porous structures with smoother surface while a several honeycombed micro-structures with fizze detached by precise tinny layers was perceived with higher in weight content of phosphorous in the residue. The larger surface area exhibited by such inter-connected system of voids alleviate

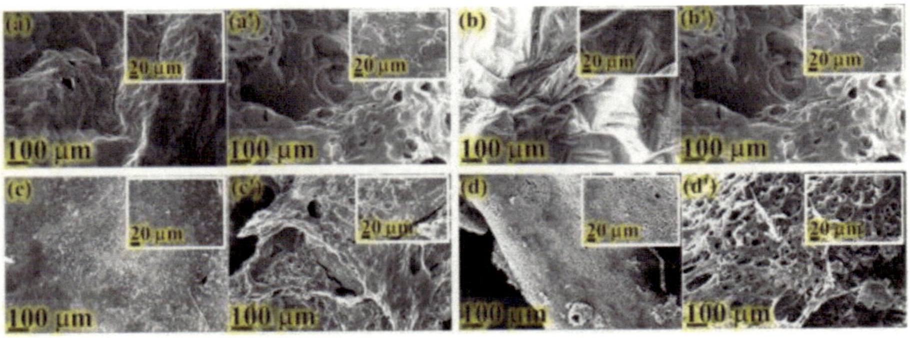

Figure 3.
SEM images of exterior (a, b, c, d) and interior (a', b', c', d') surfaces of residual char (a, a') poly(CPN_0), (b, b') poly(CPN_{10}), (c, c') poly(CPN_{80}), and (d, d') poly(CPN) samples [71].(Copyright 2018. Reproduced with permission from American Chemical Society).

altercation of heat and oxygen (air), which in turn contributes to an enhanced resistance against flame.

Further, to improve the flame resistant behaviour of polymers, the same research group have reported [72] the introduction of hexacardanolphosphazene (CP) /hexaeugenolphosphazene (EP) as a flame retardant additives in to cardanol based tris-benzoxazine monomer (C-trisapm). The flame retardant properties of resulting matrices were studied using LOI, UL-94, and smoke density analysis.

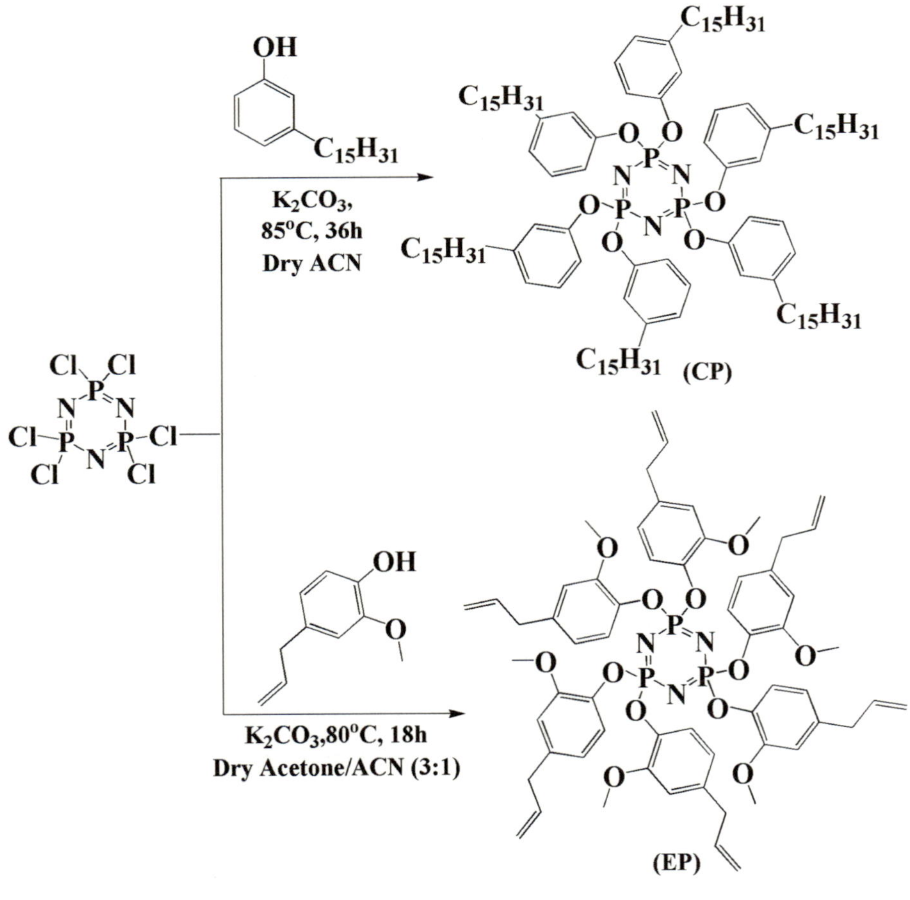

Figure 4.
Synthesis of EP and CP.

Hexaeugenolcyclotriphosphazene (EP) was synthesized (**Figure 4**) as per the reported procedure [72] as follows; in a 250 mL RB flask comprising a mixture of acetonitrile and acetone (1:3 ratio) under N_2 atmosphere, K_2CO_3 (143 mmol) was added. Eugenol (1.15 mmol) was dissolved in acetone and added to the above mixture, followed by the addition of $N_3P_3Cl_6$ (14.3 mmol). The reaction mixture was heated to 80°C and stirred for 18 h, and then allowed to cool to room temperature and evaporate the solvent. The residue was dissolved in ethyl acetate and the organic phase was washed with DM water, followed by addition of 5% NaOH and water until to obtain neutral pH. The organic layer was dried over sodium sulfate and the solvent was evaporated. Finally, the compound was purified by column chromatography using 10% ethyl acetate in hexane yield a white solid of EP.

Hexacardanolcyclotriphosphazene (CP) was synthesized (**Figure 4**) as per the reported procedure [72] using cardanol (115 mmol) was added with acetonitrile, followed by K_2CO_3 (143 mmol). A solution of $N_3P_3Cl_6$ (14 mmol) dissolved in acetonitrile was added to the reaction solution. The mixture was heated to 85°C and stirred for 36 h followed by work up as per the above procedure (EP). Finally, the crude compound was purified by column to get CP as a brown transparent liquid.

Polymer blend were prepared with three varying weight percentage of EP/CP, C-trisapm of 1:3, 1:1, and 3:1, by simple mixing of CP/EP and C-trisapm using tetrahydrofuran (THF) (**Figure 5**). After vaporizing the THF under the vacuum, the attained resultant mixture was thermally treated at 50, 100, 120,160, 180, 200, 230, and 250°C for each 1 h. The blends are abbreviated as CPxTy or EPxTy where x and y are the weight percentage ratios in the blends, and T is represented for C-trisapm.

The LOI value of poly(C-trisapm) was changed from 20 to higher values (34.58 for CP_3T_1) in the case of blends. An introduction of 1.1 wt% of phosphorous (P) in C-trisapm indicated slow burning features and all other blends exhibited self-extinguishing features. The smoke density results, C-trisapm, which burns with a smoke emission of 37.80%, while those of CP and EP with P weight percentage contents of 4.75 and 8.34 wt% exhibit the value of smoke density of 34.01 and

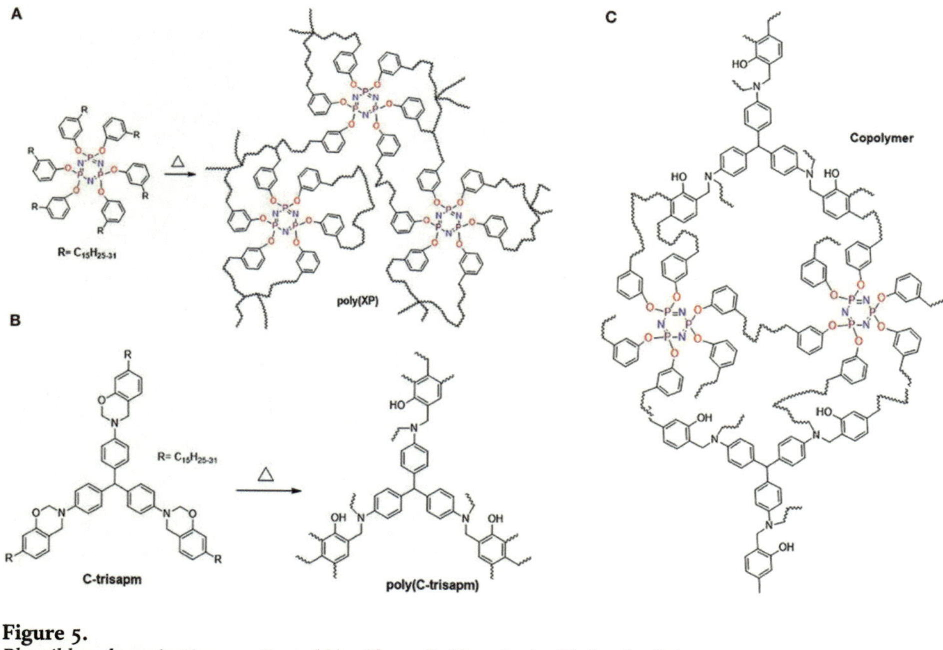

Figure 5.
Plausible polymerization reactions: (A) self-crosslinking via double bonds, (B) oxazine ring-opening polymerization, (C) Co-polymerization: Co-reaction of double bonds and oxazine ring [72]. (Copyright 2020. Reproduced with permission from Frontiers in Chemistry).

18.80 wt% respectively. Reduction in values of smoke density are due to the higher amount of phosphazene moiety present in the polymers. The UL-94 test is a vertical burning test that decides the vertical burning properties of a polymers (**Table 1**). Neat polyhexacardanolcyclotriphosphazene (poly(CP)) and its reactive blends did not catch fire instantly, dissimilar to poly(C-trisapm). The flame resistant characteristics of cross-linked CP/C-trisapm blends was found to increase with increase in P weight percent content, which supports the results of values of LOI.

Morphology of both the interior and exterior surface of the char residue attained after smoke density test was determined using SEM analysis and the images are presented in **Figure 6**. With an increasing the weight percentage of CP, the surface exterior morphology reformed to smoother surface from rippled, whereas surface interior indicated a porous honey-combed structures detached by tinny layers of border. The founding of cracks and bubbles are agreed to the stiff outer layer in case of EP comprising co-polymers, which might have rupture to discharge the interchange of heat and oxygen, thus hindering the propagation of fire as a measure of structural safety.

To improve the flame and thermal properties, M. Alagar and his research team have reported [34, 36] hybrid approach using silica (nano-silica through in-situ sol–gel/or bio silica derived from rice-husk ash) as reinforcement in to cardanol based benzoxazine (**Figure 7**). Introduction of inorganic constituent 3-mercaptopropyltrimethoxysilane (MPTMS) in to cardanol/furfurylamine based benzoxazine (BZ-C-F) via thiol-ene reaction followed by in-situ sol–gel techniques using tetraethoxysilane (TEOS). From thermal studies the percentage char yield indicates an increase of about 4.2 times (11.11 to 48.63%) and the values of LOI 21.94 to 36.95 for PBZ-C-F hybrids (**Table 2**) respectively. Functionalized bio-silica (FRHA) (0, 1, 5, 10, 15, and 20%) reinforced tri-substituted cardanol benzoxazine (CBz) (**Figure 8**), the value of LOI was increased with increase in weight percentage content of FRHA. Among the composites developed, 20 wt% FRHA reinforced CBz composites possesses the highest value of LOI of 36%, which is higher than that of values obtained for other composite samples. Also, the UL-94 vertical burning

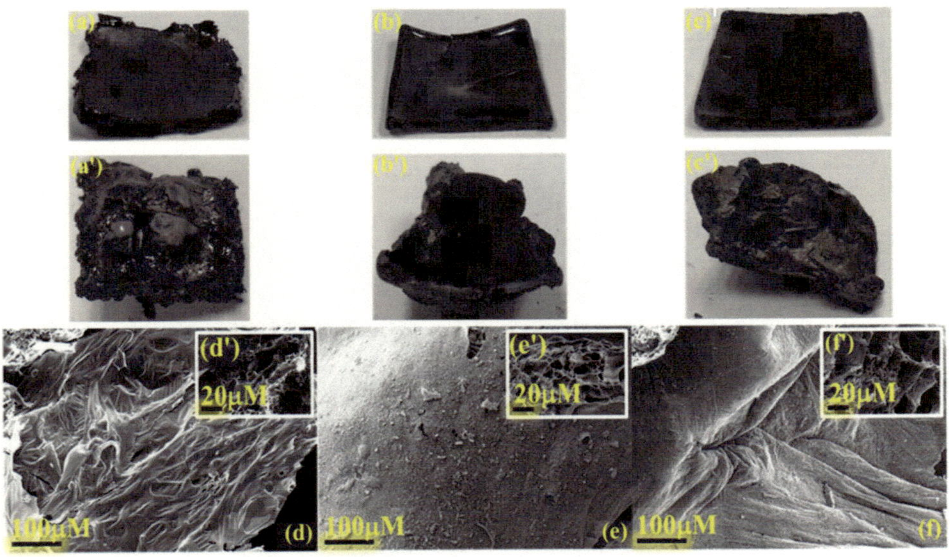

Figure 6.
Digital images of cured samples of (a) poly(CP), (b) poly(CP_3T_1), (c) poly(CP_1T_3) before (a, b, c) and after burning (a′, b′, c′); SEM images of poly(CP), poly(CP_3T_1), poly(CP_1T_3) surfaces of residual char: exterior (d, e, f) and interior (as inset) (d′, e′, f′), respectively [72]. (Copyright 2020. Reproduced with permission from Frontiers in Chemistry).

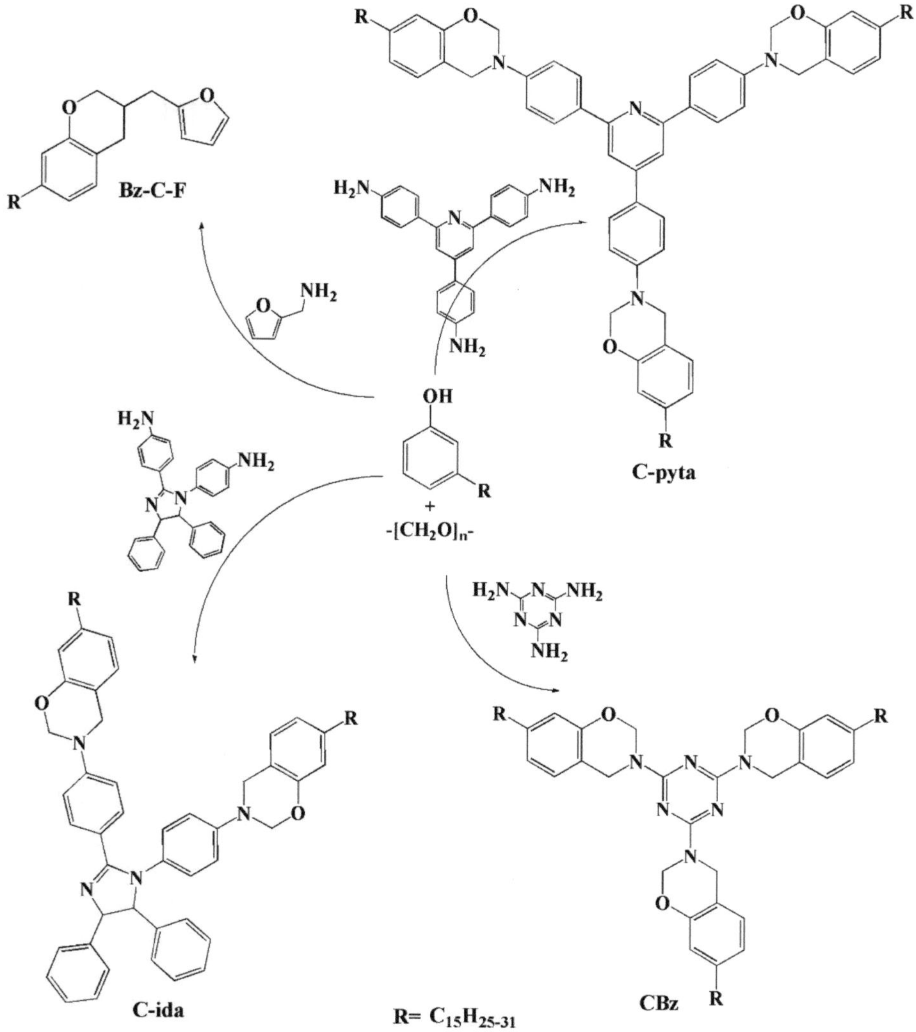

Figure 7.
Schematic synthesis of cardanol based benzoxazine monomers.

result (**Table 3**) indicates the neat CBz and CBz/FRHA (1 wt%) exhibit no rating, while 5 and 10 wt% FRHA reinforced CBz composites possess the V-2 rating. 15 and 20 wt% FRHA reinforced CBz composites showed the V-1 rating. The thermal stability and flame resistant behaviour of PBZ-silica hybrid materials are higher than those of neat PBZ as well as traditional PBA-a, due to the hybridization of silica component through chemical interaction. Further, the incorporation of higher amount of silica into the PBZ matrix reduced the volatile decomposition. In addition, the silica components offer the additional heat capacity which restricts the materials against thermal degradation.

In addition, the same research group have also reported [73] the eco-friendly cardanol-based benzoxazines (C-ida and C-pyta) synthesized from hetero-cyclic core amines (**Figure 7**), such as pyridine core triamine (pyta) and tetraarylimidazole core diamines (ida). Further, bio-composites were also prepared using 3-glycidoxypropyl-trimethoxysilane (GPTMS)-functionalized bio-silica (1, 3, 5, 7, and 10 wt%) obtained from rice husk. The thermal stability of developed PBz and its composites (**Figure 9**) is analyzed using TGA and the values obtained are presented in **Table 2**. As the bio-silica weight percentage content increases, the degradation temperature and the value

Samples	Char(%) residue	LOI	Reference
PBZ-C-F	11.11	21.94	[36]
PBZ-C-F-S-1	35.32	31.63	[36]
PBZ-C-F-S-2	41.51	34.10	[36]
PBZ-C-F-S-3	42.96	34.68	[36]
PBZ-C-F-S-4	43.77	35.01	[36]
PBZ-C-F-S-5	48.63	36.95	[36]
Poly(C-ida)	24.00	27.10	[73]
Poly(C-ida) + 1 wt% SiO_2	27.00	28.20	[73]
Poly(C-ida) + 3 wt% SiO_2	28.00	28.70	[73]
Poly(C-ida) + 5 wt% SiO_2	30.00	29.50	[73]
Poly(C-ida) + 7 wt% SiO_2	32.00	30.30	[73]
Poly(C-ida) + 10 wt% SiO_2	35.00	31.50	[73]
Poly(C-pyta)	29.00	29.10	[73]
Poly(C-pyta) + 1 wt% SiO_2	30.00	29.50	[73]
Poly(C-pyta) + 3 wt% SiO_2	32.00	30.30	[73]
Poly(C-pyta) + 5 wt% SiO_2	33.00	30.70	[73]
Poly(C-pyta) + 7 wt% SiO_2	35.00	31.50	[73]
Poly(C-pyta) + 10 wt% SiO_2	38.00	32.70	[73]
POSS-EPBz	24.00	24.00	[74]
POSS-GPBz	38.00	38.00	[74]
POSS-VPBz	36.00	36.00	[74]
SBz	20.00	25.00	[37]
SBz/FBS(1%)	21.00	26.0	[37]
SBz/FBS(3%)	23.00	27.00	[37]
SBz/FBS(5%)	25.00	28.00	[37]
SBz/FBS(10%)	27.00	29.00	[37]
FBz	46.00	36.00	[37]
FBz/FBS(1%)	47.00	36.00	[37]
FBz/FBS(3%)	48.00	37.00	[37]
FBz/FBS(5%)	50.00	38.00	[37]
FBz/FBS(10%)	53.00	39.00	[37]
PBZ-E-F	41.32	34.03	[75]
PBZ-E-F-S-1	53.64	38.96	[75]
PBZ-E-F-S-2	55.43	39.67	[75]
PBZ-E-F-S-3	62.78	42.61	[75]
PBZ-E-F-S-4	65.16	43.56	[75]
PBZ-E-F-S-5	67.54	44.52	[75]
PE-fa	52.00	38.30	[76]
PIE-fa	60.00	41.50	[76]

Samples	Char(%) residue	LOI	Reference
BGF-FPbz	52.4	38.50	[51]
BGF-SPbz	38.2	32.80	[51]

Table 2.
Thermal and flame resistant (LOI) properties of bio-based polybenzoxazine matrices and composites.

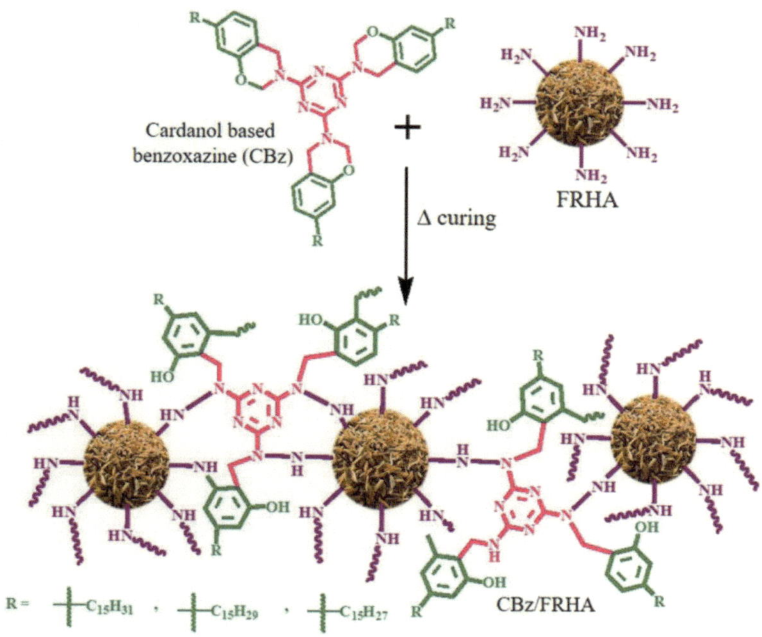

Figure 8.
The proposed schematic representation for the formation of CBz/FRHA composites [34]. (Copyright 2020. Reproduced with permission from Springer).

Samples	Char(%) residue	LOI	UL-94	Reference
CBz	12.00	22.00	—	[34]
CBz/FRHA (1 wt%)	18.00	25.00	—	[34]
CBz/FRHA (5 wt%)	32.00	30.00	V-2	[34]
CBz/FRHA (10 wt%)	36.00	32.00	V-2	[34]
CBz/FRHA (15 wt%)	40.00	34.00	V-1	[34]
CBz/FRHA (20 wt%)	47.00	36.00	V-1	[34]

Table 3.
Thermal and flame resistant (LOI and UL-94) properties of bio-based polybenzoxazine matrices and composites.

of char yield are increased. The presence of hetero-cyclic pyridine core and silica (Si-O-Si) inherently provides to the higher thermal stability. The LOI value of pyridine core cardanol-based polybenzoxazine possesses a higher residual char (29.1%) than that of other cardanol based PBz's. The values of LOI were increased with increase in weight percentage content of bio-silica. Among the composites studied, 10 wt% bio-silica reinforced composites indicated the LOI value of 32.7.

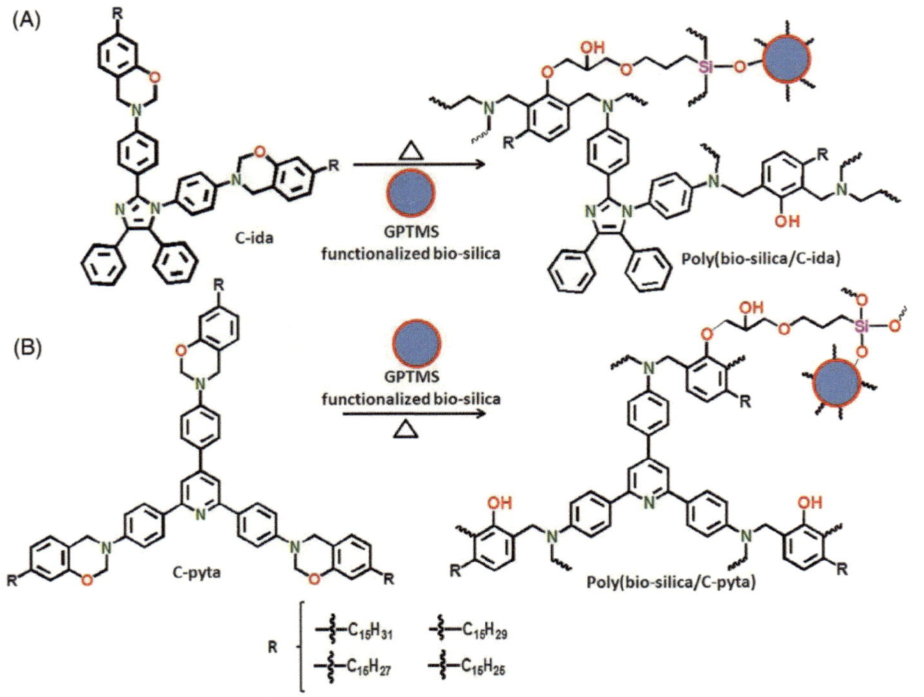

Figure 9.
Preparation of bio-silica reinforced, (A) C-ida polybenzoxazine and (B) C-pyta polybenzoxazine composites [73]. (Copyright 2020, Reproduced permission from Society of Plastics Engineers).

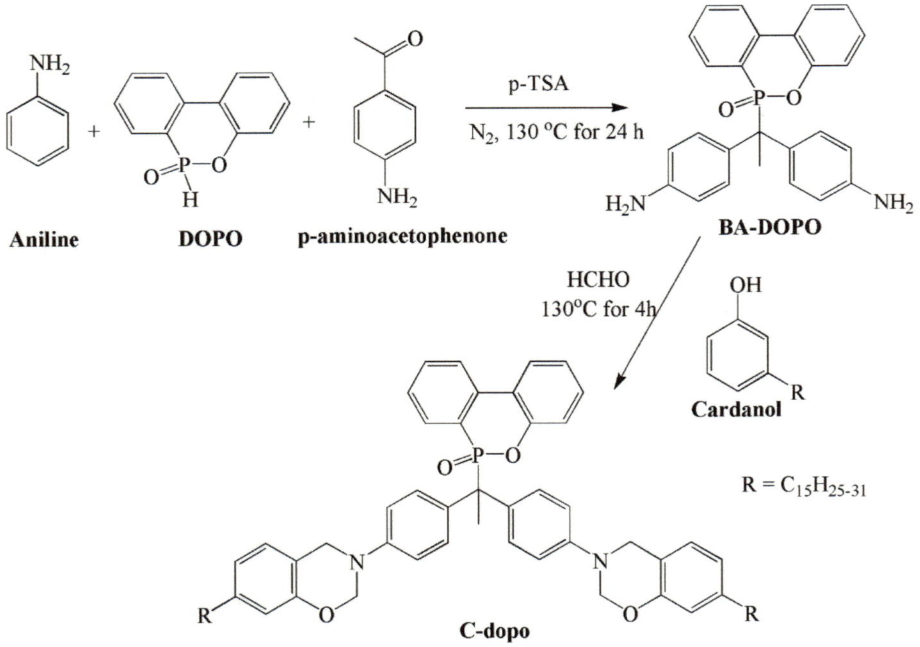

Figure 10.
Synthetic routes of BA-DOPO and cardanol based benzoxazine monomer (C-dopo).

Xin Wang and Yuan Hu research group have reported [77] the phosphorus-containing cardanol based benzoxazine (C-dopo) monomer (**Figure 10**) and was synthesized using cardanol, DOPO-based diamine (BA-DOPO) and formaldehyde.

Later, epoxy (EP)-C-dopo composites was developed by thermally activated polymerization. To addition, to improve the flame resistant efficacy, BG nano-sheets was introduced in to EP-C-dopo systems. The influence of cardanol based benzoxazine and BG nano-sheet on thermal, and flame retardant properties of EP resin was studied. The flame retardant properties of cured EP and EP/C-dopo composites were studied using LOI and UL-94 vertical burning performance, as presented in **Table 4**. Neat epoxy has a LOI value of 25%, and burns aggressively with no UL-94 classification. The introduction of flame retardant C-dopo prompts a remarkable increment in the LOI values and UL-94 rating. 10 wt% C-dopo introduced EP/C-dopo-1 system shows the LOI value of 31% and V-1 rating in UL-94 test. Further, increasing the C-dopo amount marginally increase the LOI values and achieved as V-0 rating from UL-94. In contrast, in combination of 8 wt% C-dopo and 2 wt% boron-doped graphene (BG), the EP/C-dopo/BG-1 system passed V-0 rating from UL-94 test, signifying the presence of BG effectively depresses the ignition ability of the EP composites. Further, the EP/C-dopo/BG-2 and EP/C-dopo/BG-3 systems exhibit higher LOI values and UL-94 V-0 rating performance.

The flammability behavior of cured EP and EP/C-dopo composites was further analyzed with cone calorimeter. **Table 4** shows the total heat release (THR) and heat release rate (HRR) values of EP and EP/C-dopo composites. Numerous significant flame related parameters including time to PHRR (TPHRR), time to ignition (TTI), peak heat release rate (PHRR), THR, smoke produce rate (SPR) and fire growth rate index (FIGRA) are presented in **Table 4**. From the MCC data, neat EP shows a high PHRR value of 1262 kW/m^2. While, the PHRR of the 10 wt% EP/C-dopo composites decreases to 1119 kW/m^2. Subsequent increasing of C-dopo content leads to further reduction in PHRR values to 920 and 962 kW/m^2, respectively. Further introduction of BG nano-sheet in to EP/C-dopo composites leading to the further reduction in PHRR to 870 kW/m^2, 650 kW/m^2, and 716 kW/m^2 for EP/C-dopo/BG-1, EP/C-dopo/BG-2 and EP/C-dopo/BG-3, respectively. The most substantial PHRR reduction is perceived in case of EP/C-dopo/BG-2. THR of neat EP indicates the speedy heat release and reached the value of 84.7 MJ/m^2 at the end of combustion. In contrary, both of EP/C-dopo and EP/C-dopo/BG composites observed as lower heat release to certain amount, suggesting that the introduction of C-dopo and BG into EP possibly will trigger to reduce the combustible volatile product. The incorporation of C-dopo leads to a marginal reduction in TTI. Further, the introduction of BG into EP/C-dopo increases the TTI, owing to the "tortuous path" influence of BG that retards the volatilization of combustible degradation products. Neat EP, EP/C-dopo systems reveal no changes in the value of SPR, while

Sample	Formulations			LOI %	UL-94	TTI (s)	T_{PHRR} (s)	PHRR (kW/m^2)	THR (MJ/m^2)	SPR m^2/s	FIGRA kW/m^{2}s)
	EP	CBz	BG								
EP	100	0	0	25	NR	53	110	1262	84.7	0.40	11.5
EP/C-dopo-1	90	10	0	31	V-1	49	92	1119	80.5	0.39	12.2
EP/C-dopo-2	85	15	0	32	V-0	50	95	920	79.4	0.40	9.7
EP/C-dopo-3	80	20	0	33	V-0	50	100	962	77.2	0.38	9.6
EP/C-dopo/BG-1	90	8	2	30	V-0	49	89	870	75.9	0.29	9.8
EP/C-dopo/BG-2	85	13	2	33	V-0	52	120	650	74.4	0.29	5.4
EP/C-dopo/BG-3	80	18	2	33	V-0	56	120	716	78.7	0.31	6.0

Table 4.
Formulations and flame retardant properties of neat EP, EP/CBz and EP/CBz/BG composites.

the EP/C-dopo/BG system indicate the lower value of SPR, designating that BG nano-sheet with proficient barricade effect can help as suppressant of smoke in the burning method. The FIGRA of the neat EP is 11.5 kW/(m^2·s). In comparison, excluding that the EP/C-dopo-1 system displays a marginally improved value of FIGRA, all other systems display lower value of FIGRA. The most noticeable result is perceived in case of EP/C-dopo/BG-2 (5.4 kW/(m^2s)), suggesting the considerably inhibited fire hazards.

3. Eugenol based polybenzoxazines for flame resistant applications

Among renewable phenolic alternatives to substitute the conventional phenolic, eugenol appears as most imminent raw-materials for the sustainable PBz's production owing to their abundance availability and low cast production. Eugenol is obtained from several bio-sources including, clove, tulsi, cinnamon, pepper, turmeric, and thyme. The eugenol structure is very exciting as allyl functional group permits additional cross-linking with numerous functional groups. In addition, the phenolic group offers numerous chemical reactions and formation of products including esterification, cyanogenation, alkylation, and novalac, epoxy and BZs, etc.

Sarojadevi Muthusamy research group reported [74] the successful synthesis of the polyhedral oligomeric silsesquioxane (POSS) tethered PBz nano-composites using renewable bio-phenols (eugenol, guaiacol, vanillin), POSS-octaamine with paraformaldehyde through Mannich reaction. The developed POSS-PBz nanocomposites (**Figure 11**) from sustainable bio-sources have a great potential application as high-performance materials owing to its excellent thermal and flame resistant properties. Thermal and flame resistant properties of POSS-PBz nanocomposites were determined using TGA under nitrogen atmosphere. The char yield percent and LOI values (**Table 2**) are found to be 24, 38, 36 and 27.1, 32.7, 31.9 respectively for POSS-EPBz, POSS-GPBz, and POSS-VPBz. Among the nanocomposites developed POSS-GPBz nanocomposites shows better thermal and flame resistant properties over other two POSS-PBz systems.

Alagar group have also reported [37] the new type of bio-based PBz's synthesized using eugenol and furfurylamine/stearylamine through greener synthetic

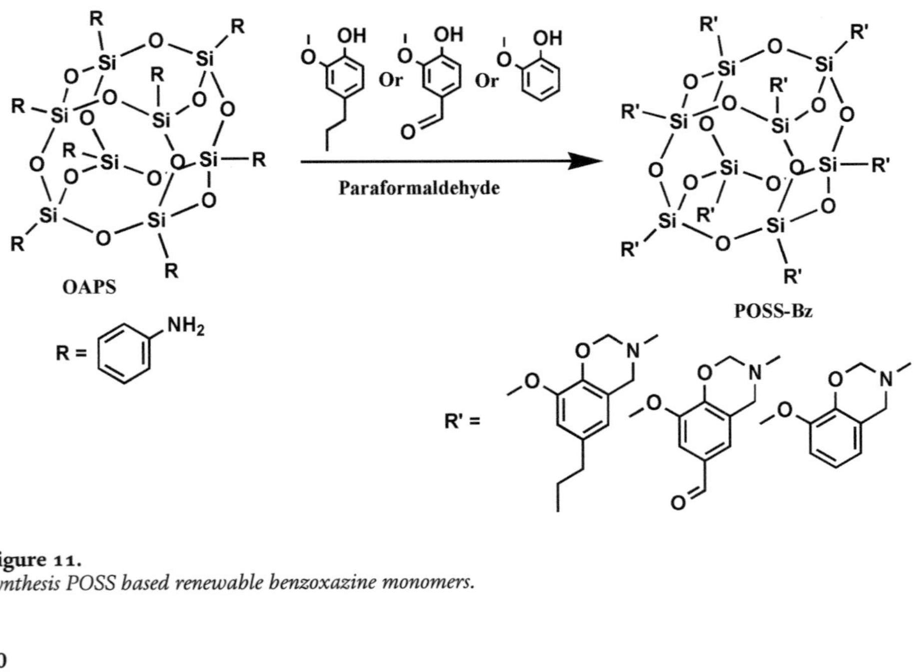

Figure 11.
Synthesis POSS based renewable benzoxazine monomers.

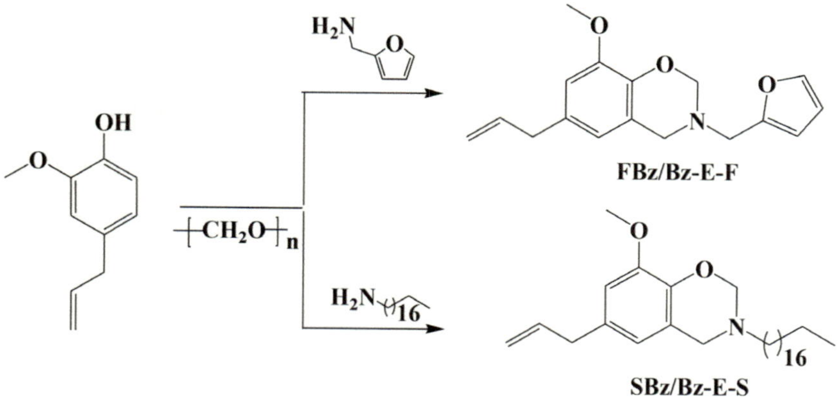

Figure 12.
Synthesis of eugenol based benzoxazine monomers using different amines.

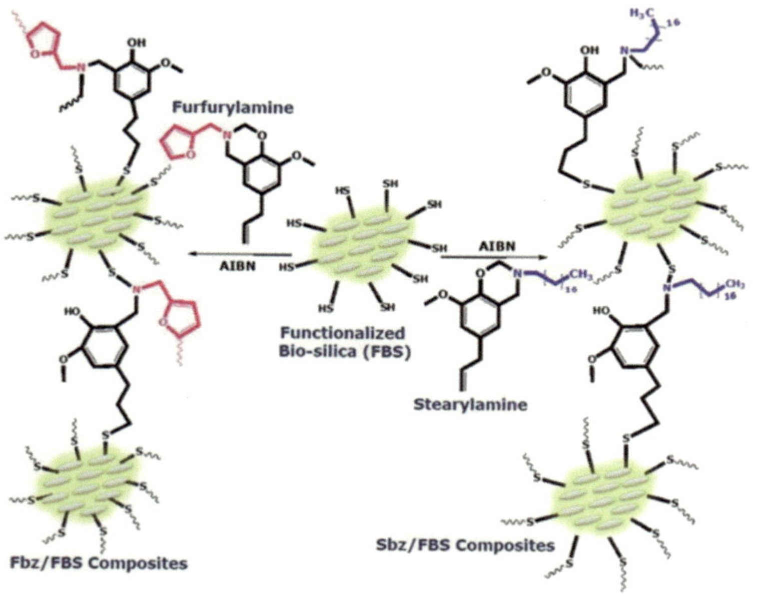

Figure 13.
The schematic representation for the formation of FBS/FBz and FBS/SBz [37] composites.

approach (**Figure 12**). The synthesized benzoxazine monomers were further reinforced with varying weight percentage content (0, 1, 3, 5, and 10 wt%) of bio-silica obtained from rice-husk to fabricate PBz hybrid composites (**Figure 13**). The TGA results obtained inferred that the developed bio-silica reinforced PBz composites possess the higher thermal and improved flame resistant properties. The furfuryl amine based PBz (FBz), the char yield value was increased to 53% from 46%. In case of stearyl amine based PBz (SBz), the char yield value was increased to 27% from 20%. The value of char yield of SBz is lower than that of FBz because of the presence of long aliphatic chain. In common, the polymeric materials with LOI values more than 26 are considered as flame-retardant materials, it was found that the bio-silica reinforced bio-based PBz's composites possess the LOI values of 25 to 39 (**Table 2**), signifying an enhanced flame resistant nature of the developed bio-based PBz composites.

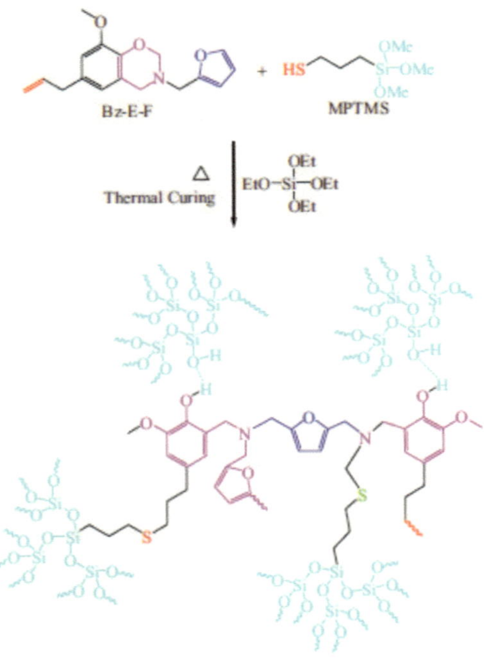

Figure 14.
Schematic representation for the preparation of renewable PBZ–E–F– silica hybrids [75].

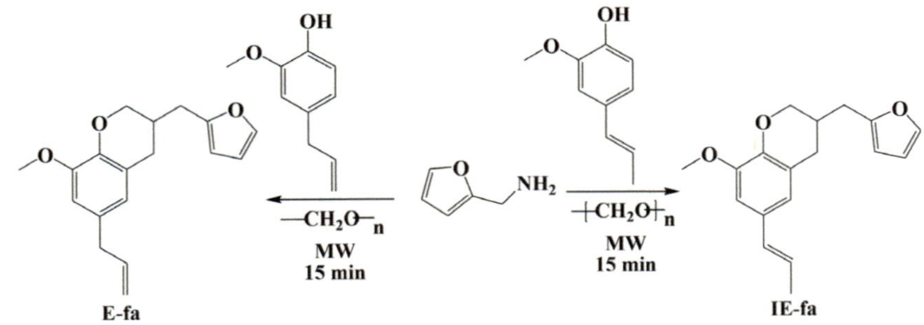

Figure 15.
Synthesis of renewable bio-based isomeric benzoxazine E-fa and IE-fa.

In addition, the same group have further reported [75] the high thermal and flame resistant PBz-silica hybrid materials were synthesized using eugenol and furfurylamine via sol–gel-techniques (**Figure 14**). An inorganic component of TEOS was introduced into eugenol- benzoxazine (BZ–E–F) with the help of MPTMS as a coupling agent viz. thiol-ene click approach. The thermal studies indicate that the char yield increases to 67.54 from 41.32 (neat PBz) and LOI increased to 44.52 for PBZ–E–F silica hybrid from 34.03 for neat PBz (**Table 2**).

Bimlesh Lochab research group have reported [76] the synthesis of sustainable bio-based benzoxazine monomers (**Figure 15**) using isomeric phenols, eugenol (E) and iso-eugenol (IE), and furfurylamine (fa) to form E-fa and IE-fa monomer, respectively. The monomers differ in the position of the double bond in the p-substituted propylene unit forming non-conjugated Vs. conjugated alkylene chain with the benzene ring containing benzoxazine in E-fa and IE-fa respectively. In comparison to other bio-based PBz's, both PE-fa and PIE-fa exhibited the higher thermal stability and high flame resistant properties. The char residue values of PE-fa and PIE-fa are 52 and 60 respectively and the LOI values observed are 38.3 and 41.5 respectively (**Table 2**).

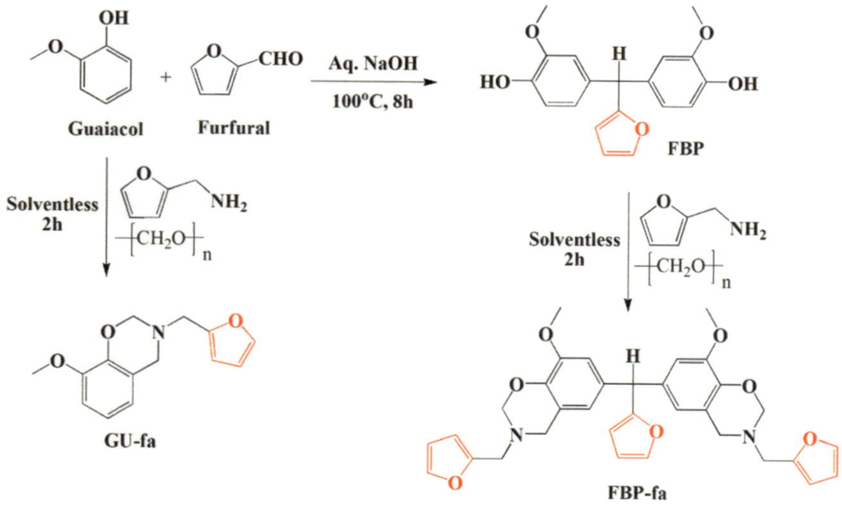

Figure 16.
Synthesis of fully bio-based mono and tri-furan functional mono and bis-benzoxazine resin.

4. Guaiacol based polybenzoxazine for flame resistant applications

Kan Zhang research group have developed [78] the fully bio-based tri-furan functional bis-benzoxazine resin (**Figure 16**) and was synthesized using furfural, guaiacol, furfurylamine and paraformaldehyde via a two-step reaction approach. In the first step, bio-based bis-phenol containing furan ring was synthesized using furfural and guaiacol via base mediated condensation reaction. Then the bio-based sustainable tri-furan functional bis-benzoxazine (FBP-fa) was synthesized using bio-based bis-phenol, furfurylamine and paraformaldehyde through greener approach. The flammability behaviour of poly(GU-fa) and poly(FBP-fa) was assessed using LOI. As a result, poly(GU-fa) and poly(FBP-fa) show LOI values of 39.9 and 42.3 respectively. Both PBz's possess better LOI values in the self-extinguishing region (LOI > 28). Also MCC analysis was utilized further to assess the flammability performance of poly(GU-fa) and poly(FBP-fa). From the MCC characterization of poly(GU-fa) and poly (FBP-fa) display HRC values of 79.6 and 30.4 J/gK, respectively (**Table 5**).

Sample	Char residue (%)	LOI %	HRC (J/gK)	THR (KJ/g)	Reference
Poly(GU-fa)	56	39.9	70.6	6.5	[78]
Poly(FBP-fa)	62	42.3	30.4	5.8	[78]
Poly(RES-a)	55		79.7	15.9	[45]
Poly(RES-ac)	74		30.7	6.0	[45]
Poly(RES-ch)	59		73.4	14.2	[45]
Poly(RES-fa)	64		54.0	9.3	[48]
Poly(API-fa)-1	63	42.7	22.5	11.2	[79]
Poly(API-fa)-1	66	43.9	20.2	9.4	[79]
Poly(NAR-fa)	64		31.9	6.6	[80]

Table 5.
Formulations and flame retardant properties of bio-based polybenzoxazines.

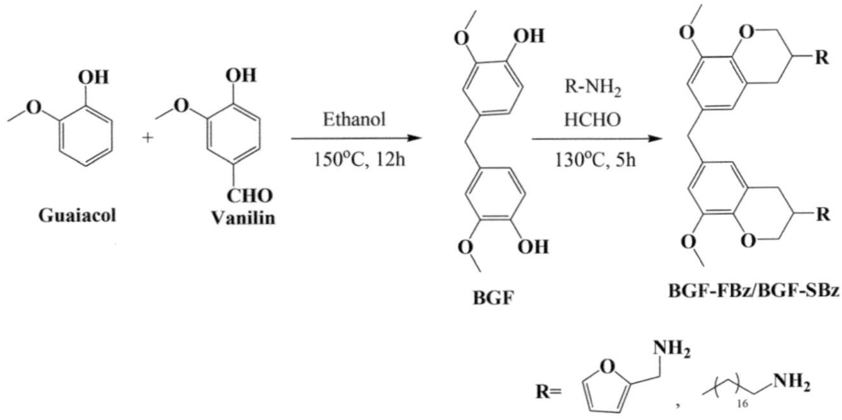

Figure 17.
Synthesis of bisguaiacol-F (BGF) and BGF based benzoxazine monomers.

Further, poly(GU-fa) infers the THR value of 6.5 KJ/g, whereas poly(FBP-fa) displays a fairly lower THR value of 5.8 KJ/g. The data of flame related properties of both PBz's are presented in **Table 5**. Unexpectedly, the HRC value of developed poly(FBP-fa) gives significantly lower value than that of earlier reported polymers. Furthermore, poly(FBP-fa) displays very lower flammability characteristics than that of many other reported PBz's. It is well-known that the HRC value less than 300 J/gK can be observed as self-extinguishing while, whereas the HRC values less than 100 J/gK are designated as non-ignitable materials.

M. Sarojadevi research group have also synthesized [51] the BGF and BGF-based benzoxazines (**Figure 17**) viz., bis[(3-(furan-2-yl) methyl)-3,4-dihydro-2H benzoxazin-6-yl] methane and bis[(3-octadecyl)-3,4-dihydro-2H benzoxazin-6-yl] methane as an substitute to BPA. The LOI values (**Table 2**) of the BGF-FPbz and BGF-SPbz was found to be 38.5 and 32.8 respectively indicating the high flame retardancy. As expected, the developed PBz's with these LOI values are greater than 26 endorsing outstanding flame resistant property.

5. Resveratrol based polybenzoxazine for flame resistant applications

Hatsuo Ishida and Kan Zhang group reported [45, 48] the series of thermally stable and flame resistant resveratrol-based tri-functional benzoxazine monomers

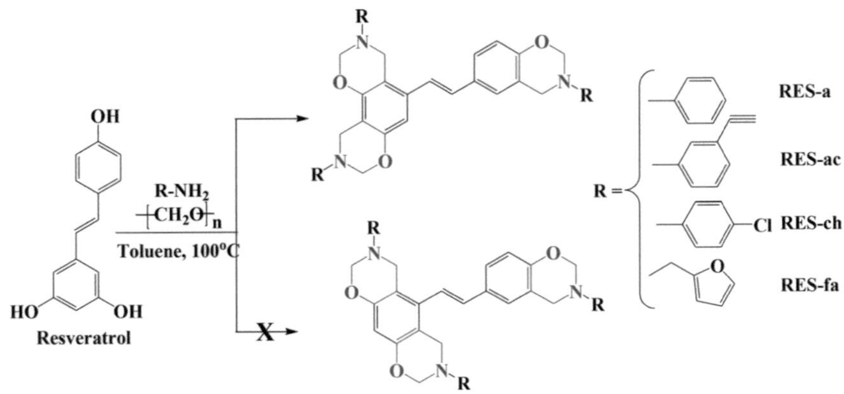

Figure 18.
Synthesis of resveratrol-based tri-functional benzoxazine monomers.

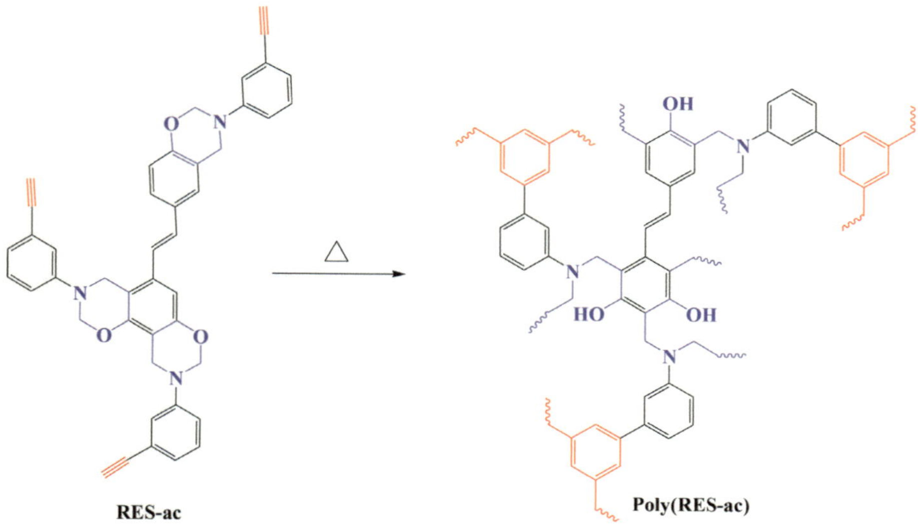

Figure 19.
Schematic representation of proposed thermal behaviors of RES-ac.

(**Figure 18**) have been produced using resveratrol, different amines (aniline, 4-chloroaniline, 3-aminophenylacetylene, and furfurylamine) and paraformaldehyde. The thermal and flammability performances of the developed PBz's are studied using TGA and MCC respectively. Among the benzoxazine studied, resveratrol-based, tri-functional benzoxazines containing acetylene possesses the highest thermal stability with a value of T_g over the 350°C, 10% weight loss at 465°C, and char yield of 74%. Further, the developed polybenzoxazine infers the extremely low heat release capacity of 30.7 J/gK and total heat release value of 6.0 kJ/g.

MCC analysis was performed to analyse the quantitative features of flame resistance of PBz's. Usually, the extreme value of specific HRR can be used to assess the HRC, which is considered as unique best single interpreters for the flame resistance of materials. Here, poly(RES-a), poly(RES-ch), poly(RES-ac) (**Figure 19**), and poly (RES-fa) were analysed at the heating rate of 1°C/s over the temperature range 100 ~ 750°C. As illustrated in MCC results (**Table 5**) of poly(RES-a), poly(RES-ch), poly(RES-ac) and poly(RES-fa) indicate HRC values of 79.7, 73.4, 30.7, and 54 J/gK, respectively. Moreover poly(RES-a), poly(RES-ch), and poly(RES-fa) exhibit THR values of 15.9, 14.2, and 9.3 KJ/g, while poly(RES-ac) shows a lower THR value of 6.0 KJ/g. The data obtained from thermal and flame resistant properties for PBz's are presented in **Table 5**. The value of HRC for poly(RES-ac) is significantly lesser than those of other polymers. Moreover, the HRC value of poly (RES-ac) is even lower than that of the thermosets from ortho-amide and ortho-imide benzoxazines, which exhibited one of the lowest HRC values of all polymers. Obviously, the PBz formed from the ring opening polymerization along with the cyclo trimerization of acetylene results in a considerable HRC reduction. Thus, the newly developed resveratrol-based tri-benzoxazine monomer containing acetylene group has greatest prospective applications as anti-flammable and fire resistant matrix for advanced composites.

6. Apigenin based polybenzoxazine for flame retardant applications

Kan Zhang research group reported [79] the synthesis of a new fully bio-based bis-benzoxazine (API-fa) (**Figure 20**) using apigenin as phenolic precursor and

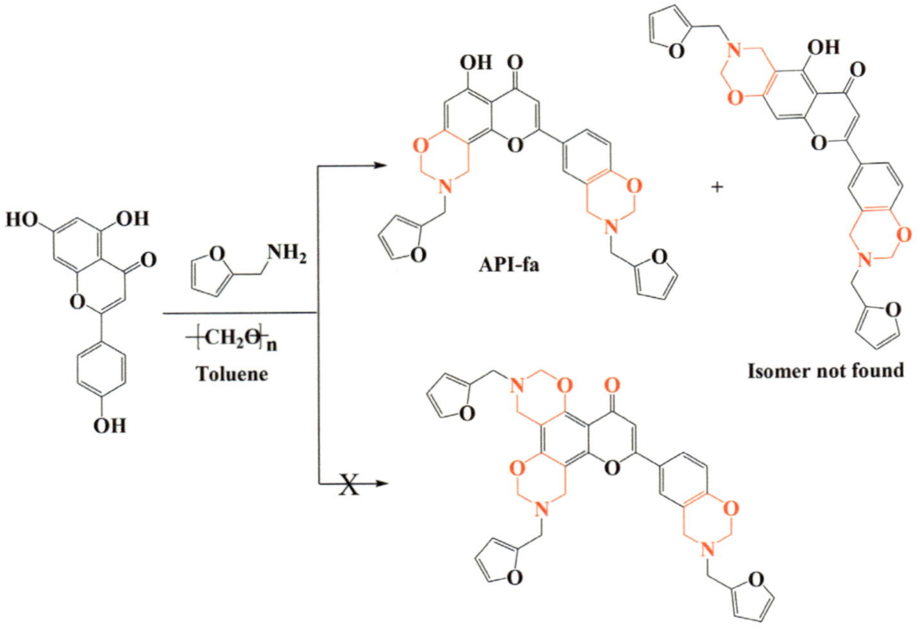

Figure 20.
Synthesis of apigenin-based bis-benzoxazine (API-fa).

furfurylamine as amine precursor both derived from bio-sources. The well-structured intra-molecular hydrogen bond in API-fa is probable to be a stable latent heat formulation, which prolongs the shelf life of the Bz resin. Besides, the benzopyrone and furan rings with carbon–carbon double bonds in the API-fa are proficient of making further cross-linking networks, which are anticipated to considerably increase the mechanical, thermal and flame resistant properties of the resulting PBz thermosets. Remarkably, the thermal possessions of the API-fa based PBz achieved in this work are considerably higher than those of recently reported PBz's. The resulting polymerized PBz thermoset developed from API-fa indicates the high glass transition temperature of 376°C, a higher char residue of 66%, with an extremely lower heat release capacity of 20.2 J/gK (**Table 5**).

Thus, the LOI values obtained for poly(API-fa)-1 and poly(API-fa)-2 are as high as 42.7 and 43.9, respectively (**Table 5**). From the MCC analysis of poly(API-fa)-1

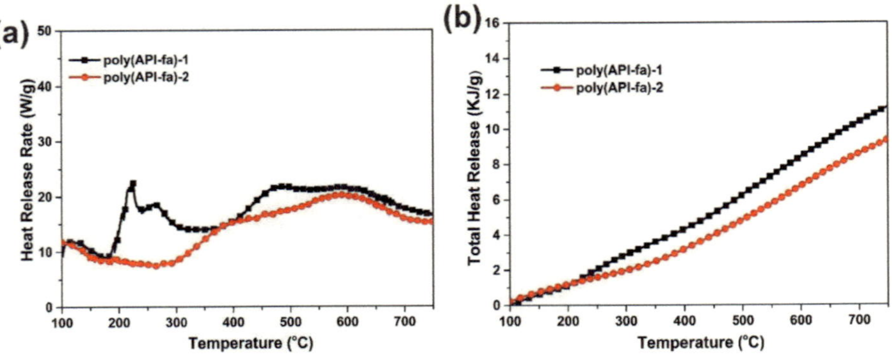

Figure 21.
Microscale combustion calorimetric (MCC) analysis of poly(API-fa)-1 and poly(API-fa)-2. (a) heat release rate, and (b) total heat release as a function of the temperature [79]. (Copyright 2020, reproduced permission from Royal Society of Chemistry).

and poly(API-fa)- 2 was presented in **Figure 21**. **Figure 21a**, HRR maximums are perceived at 225°C and 591°C for poly(API-fa)-1 and poly(API-fa)-2, respectively. Also, the values of HRC are observed as 22.5 and 20.2 J/gK respectively. Furthermore, poly(API-fa)-1 indicates the THR value of 11.2 KJ/g, whereas poly(API-fa)-2 shows a moderately lower value of THR (9.4 KJ/g) (**Figure 21b**). These results support the very exceptional flame retardant behaviour of apigenin based polybenzoxazines.

7. Naringenin based benzoxazine for flame retardant applications

Kan Zhang et al., synthesized [80] a fully bio-based benzoxazine monomer (NAR-fa) using Naringenin, furfurylamine and paraformaldehyde (**Figure 22**). Thermal properties of the resulting co-polymeric thermosets were maintained or slightly enhanced, while those related to flame retardancy improved to about 38 and 51% for THR and HRC, respectively (**Table 5**).

Figure 23 displays the graphic representation of the HRR as a function of temperature, where the maximum HRR is detected at 455°C. From **Figure 23**, it was calculated a HRC and THR of 31.9 J/gK and 6.6 KJ/g, respectively. The HRC value for poly(NAR-fa) is significantly lower than those reported for PBz's with high

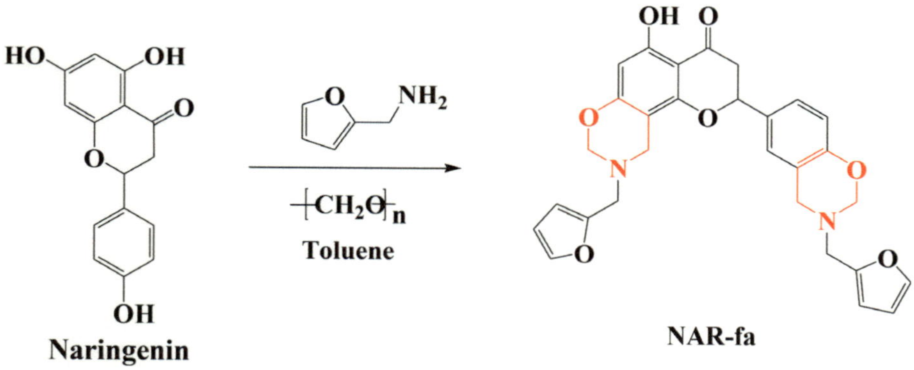

Figure 22.
Synthesis of fully bio-based naringenin-benzoxazine monomer (NAR-fa).

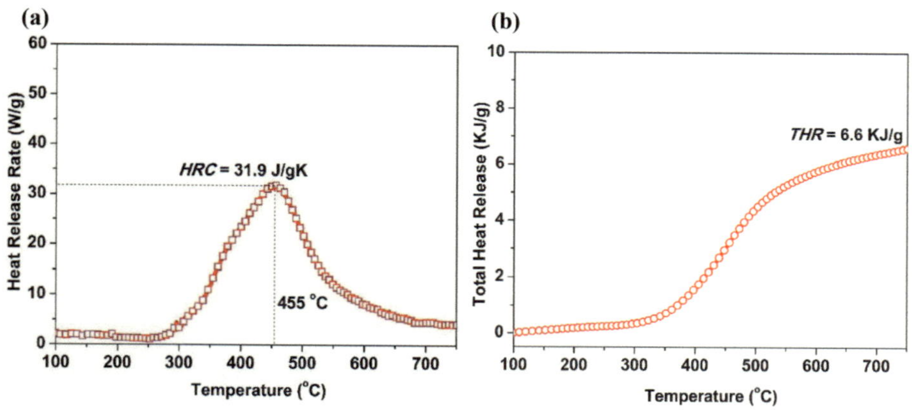

Figure 23.
Microscale combustion calorimetric (MCC) analysis of poly(NAR-fa). (a) heat release rate, and (b) total heat release as a function of the temperature [80]. (Copyright 2020, reproduced permission from Royal Society of Chemistry).

flame resistance properties attained upon polymerization of ortho-imide and ortho-amide functionalized benzoxazine monomers. These results show that poly (NAR-fa) can indeed be classified as non-ignitable given that its HRC is lower than 100 J/gK.

8. Phloroglucinol based polybenzoxazine for flame retardant applications

Phloroglucinol (P) is also a kind of bio-based renewable phenolic material widely occurred in greenery plants, and it contains three phenolic hydroxyl groups can form a three-arm benzoxazine monomer, which possibly increase the cross-linking density of cured PBz resin result in higher thermal stability, mechanical behaviour and other properties. Zuomin Zhan et al. reported [81] the development of fully bio-based low temperature cured polybenzoxazines. In order to achieve high thermal stability, bio-mass-derived phloroglucinol (P) as phenol source, furfurylamine (FA) or P-amino benzoic acid (PABA) as amine source and paraformaldehyde were used to synthesize two novel fully bio-based benzoxazines (**Figure 24**). The obtained fully bio-based polybenzoxazines possesses a high char yield of 53.0%, low HRC of 37 J/gK and a low THR of 8.3 kJ/g.

The key combustion parameters, including HRC, THR, PHRR and TPHRR, were obtained and summarized in **Table 6**. MCC curves of all system exhibited dissimilar two heat release peaks. HRC, THR and PHRRs of P(P-paba) were 86 J/gK, 12.3 kJ/g, 79.6 W/g and 39.6 W/g respectively and only about half of P(BA-a) (169.0 J/g·K, 23.6 kJ/g, 166.6 W/g and 94.0 W/g). Its TPHRR2 (528°C) was 9°C higher than that of P(BA-a) (519°C). These consequences were ascertained that the formation of H-bonding in the molecular structure of P(P-paba) might effectively diminish the

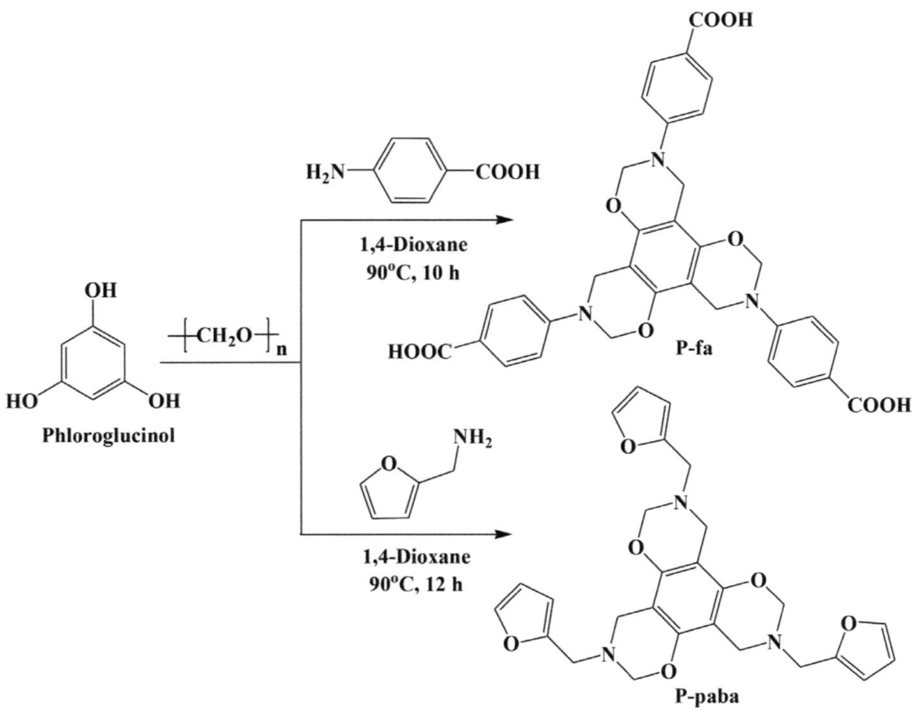

Figure 24.
Synthesis phloroglucinol based benzoxazine monomers (P-fa and P-paba).

Sample	Char residue (%)	HRC (J/gK)	THR (KJ/g)	$PHRR_1$ (W/g)	T_{PHRR1} (°C)	$PHRR_2$ (W/g)	T_{PHRR2} (°C)	Reference
P(P-fa)	53	37	8.3	37.8	348	21.8	441	[81]
P(P-fa (9.5) + P-paba (0.5))	52	39	9.4	34.5	323	22.9	441	[81]
P(P-fa (9) + P-paba (1))	53.3	38	9.2	34.0	312	22.3	454	[81]
P(P-fa (8) + P-paba (2))	50.3	45	9.8	39.7	305	23	465	[81]
P(P-paba)	34.7	86	12.3	79.6	300	39.6	528	[81]
P(BA-a)	30.6	169	23.6	166.6	448	94.0	519	[81]

Table 6.
Thermal and flame retardant properties of bio-based phloroglucinolbased tri-functional polybenzoxazines.

heat release values. Though, TPHRR1 of P(Ppaba) was about 300°C, signifying again that the degradation of benzoxazine diminished without Mannich structure formation. In comparison with P(P-paba), HRC value of 37 J/gK, THR value of 8.3 kJ/g and PHRRs value of 37.75 W/g and 21.84 W/g of P(P-fa) diminished considerably, and merely 20–35% of those of P(BA-a), specifying that the presence of furan group could enhance the additional cross-linking density and greatly decrease the heat release of fully bio-based sustainable benzoxazine resins. Thus, it may possibly be expected that P(P-paba) and P(P-fa) had possesses best flame resistant behaviour than that of P(BA-a).

9. Summary and conclusion

In the present chapter, we have reviewed the synthesis of bio-based benzoxazine monomers with varying molecular design including mono, di, tri, tetra, hexa, and octa functional oxazine using renewable bio based raw materials such as cardanol, eugenol, guaiacol, vanillin, phloroglucinol, naringenin, apigenin, resveratrol, furfurylamine, stearylamine and etc. The present chapter also discusses the preparation of polybenzoxazine bio-composites using bio-silica, silica particle through sol–gel approach and boron complexes and their thermal and flame resistant properties using various characterization techniques including LOI, UL-94 vertical flame test, smoke density, SEM analysis, and MCC. Data obtained from the thermal and flame resistant analysis, the newly developed resveratrol, phloroglucinol, naringenin, apigenin based multi-functional benzoxazine has greatest prospective applications as anti-flammable and fire resistant matrix for advanced composites for several applications. Especially, the HRC and TRC value of poly(RES-ac) is ominously lower than that of the thermosets from ortho-amide and ortho-imide benzoxazines, which exhibited one of the lowest HRC values of all polymers. Thus, this newly developed resveratrol-based tri-benzoxazine monomer containing acetylene group has greatest prospective applications as anti-flammable and fire resistant matrix for advanced composites. Consequently, these developments might be very useful in fabricating new products for the next generation's high-performance and flame retardant uses by exploiting eco-friendly, sustainable and cost competitive bio-based polybenzoxazine matrices and composites for varying range of industrial and engineering applications.

Author details

Subramani Devaraju[1]*, Krishnamoorthy Krishnadevi[1] and Muthukaruppan Alagar[2]*

1 Polymer Composites Lab, Division of Chemistry, Department of Science and Humanities, Vignan's Foundation for Science, Technology and Research (VFSTR), Guntur, India

2 Polymer Engineering Laboratory, PSG Institute of Technology and Applied Research, Coimbatore, India

*Address all correspondence to: subudeva@gmail.com
and muthukaruppanalagar@gmail.com

References

[1] Zhao P, Development and investigation of bio-based environmentally friendly fire retardant PLA composites, Dissertation at the Faculty of Mechanical E ngineering, Chemnitz University of Technology, Professorship of Plastic Materials, Chemnitz, 2018.

[2] https://firesafeeurope.eu/wp-content/uploads/2018/09/Europe-is-playing-with-fire-_-White-Paper_FSEU_03_12_2014.pdf

[3] McKenna ST, Birtles R, Dickens K, Walker RG, Spearpoint MJ, Stec AA, Hull TR, Flame retardants in UK furniture increase smoke toxicity more than they reduce fire growth rate. Chemosphere. 2018;196:429–439.

[4] Dewaghe C, Lew CY, Claes M, Belgium SA, Dubois P, Fire-retardant applications of polymer–carbon nanotubes composites: Improved barrier effect and synergism. In Polymer–Carbon Nanotube Composites; McNally T, Pötschke P, (Eds.), Woodhead Publishing: Cambridge, UK, 2011; 718–745.

[5] Kim Y, Lee S, Yoon H, Fire-Safe Polymer Composites: Flame-Retardant Effect of Nanofillers. Polymers. 2021;13: 540.

[6] Horrocks, AR, Price D, (Eds.), Advances in Fire Retardant Materials, Woodhead Publishing: Cambridge, UK, 2008.

[7] Nair CPR, Advances in addition-cure phenolic resins, Progress in Polymer Sciences. 2004;29:401-498.

[8] Ohashi S, Kilbane J, Heyl T, Ishida H, Synthesis and characterization of cyanate ester functional benzoxazine and its polymer, Macromolecules. 2015; 48:8412-8417.

[9] Auvergne R, Caillol S, David G, Boutevin B, Biobased thermosetting epoxy: present and future, Chemical Review. 2014;114:1082-1115.

[10] Gandini A, Polymers from Renewable Resources: A challenge for the future of macromolecular material, Macromolecules. 2008;41:9491-9504.

[11] Liu Y, Cao L, Luo J, Peng Y, Ji Q, Dai J, Zhu J, Liu X, Biobased nitrogen-andoxygen-codoped carbon materials for high-performance supercapacitor, ACS Sustainable Chemistry and Engineering. 2018;7:2763-2773.

[12] Dai Y, Teng N, Shen X, Liu Y, Cao L, Zhu J, Liu X, Syntheis of biobased benzoxazines suitable for vacuum-assisted resin transfer molding process viaintroduction of soft silicon segment, Industrial & Engineering Chemistry Research. 2018;57:3091-3102.

[13] Thomas S, Joseph K, Malhotra SK, Goda K, Sreekala MS, (Eds.), Polymer Composites: Volume 1 Macro and Microcomposites. Wiley, 2012.

[14] Njuguna J, Pielichowski K. Polymer nanocomposites for aerospace applications: properties. Advanced Engineering Materials. 2003;5:769-78.

[15] Devaraju S, Alagar M. Unsaturated Polyester – Macrocomposites, 2018. Elsevier

[16] Devaraju S, Vengatesan MR, Selvi M, Ashokkumar A, Hamerton I, Alagar M, low surface free energy cyanate ester-silica hybrid (CE-SiO_2)nanomaterials for low k dielectric applications, RSC Advances. 2013;3:12915–12921.

[17] John G, Nagarajan S, Vemula PK, Silverman JR, Pillai CKS, Natural Monomers: A Mine for Functional and Sustainable Materials - Occurrence, Chemical Modification and Polymerization, Progress in Polymer Sciences. 2019;92:158-209.

[18] Iwata T, Biodegradable and bio-based polymers: future prospects of eco-friendly plastics. Angewandte Chemie International Edition. 2015;54:3210–3215.

[19] Voirin C, Caillol S, Sadavarte N V, Tawade B V, Boutevin B, Wadgaonkar PP, Functionalization of cardanol: towards biobased polymers and additives, Polymer Chemistry. 2014;5:3142-3162.

[20] Lora JH, Glasser WG, Recent industrial applications of lignin: A sustainable alternative to nonrenewable materials. Journal of Polymers and the Environment. 2002;10:39–48.

[21] Ramon E, Sguazzo C, Moreira PMGP, A Review of Recent Research on Bio-Based Epoxy Systems for Engineering Applications and Potentialities in the Aviation Sector. Aerospace. 2018;5:110.

[22] Akampumuza O, Wambua PM, Ahmed A, Li W, Qin X-H, Review of the applications of biocomposites in the automotive industry, Polymer composites. 2017;38:2553-2569.

[23] Antunes A, Faria P, Silva V, Brás A, Rice husk-earth based composites: A novel bio-based panel for buildings refurbishment, Construction and Building Materials. 2019; 221:99-108.

[24] Ishida H, Froimowicz P, Advanced and Emerging Polybenzoxazine Science and Technology, John Fedor publisher, Elsevier, 2017.

[25] Ishida H, Agag T, Handbook of Benzoxazine Resins, Elsevier, Amsterdam, 2011.

[26] Kiskan B, Adapting Benzoxazine Chemistry for Unconventional Applications, Reactive and Functional Polymers. 2018;129:76-88.

[27] Ghosh NN, Kiskan B, Yagci Y. Polybenzoxazines - New high performance thermosetting resins: Synthesis and properties, Progress in Polymer Sciences. 2007;32:1344-1391.

[28] Vengatesan MR, Devaraju S, Dinakaran K, Alagar M. SBA-15 filled polybenzoxazine nanocomposites for low-k dielectric applications, J. Mater. Chem. 2012, 22, 7559-7566.

[29] Yagci Y, Kiskan B, Ghosh NN, Recent advancement on polybenzoxazine-A newly developed high performance thermoset, Journal of Polymer Sciences. Part-A: Polymer Chemistry. 2009;47:5565-5576.

[30] Selvi M, Devaraju S, Vengatesan MR, Kumar M, Go JS, Alagar M. The effect of UV radiationon Polybenzoxazine/epoxy/OG-POSS nanocomposites, RSC Advances. 2014;4:8238-8244.

[31] Selvi M, Vengatesan MR, Devaraju S, Kumar M, Alagar M. In-situ sol-gel synthesis of silica reinforced polybenzoxazine hybrid materials with low surface free energy, RSC Advances. 2014;4:8446-8452.

[32] Selvi M, Devaraju S, Alagar M. Cyclotriphosphazene nanofiber-reinforced polybenzoxazine/epoxy nanocomposites for low dielectric and flame-retardant applications, Polymer Bulletin. 2019;76:3785–3801.

[33] Hariharan A, Srinivasan K, Murthy C, Alagar M. A Novel Imidazole-Core-Based Benzoxazine and Its Blends for High-Performance Applications, Industrial & Engineering Chemistry Research. 2017;56:9347–9354.

[34] Krishnadevi K, Devaraju S, Naveena E, Alagar M, Environmentally sustainable rice husk ash reinforced cardanol based polybenzoxazine bio-composites for insulation applications, Polymer Bulletin. 2020;77:2501-2520.

[35] Devaraju S, Krishnadevi K, Naveena E, Alagar M, Design and

Development of Polybenzoxazine-POSS Hybrid Materials from Renewable Starting Materials for Low K and Low Surface Free Energy Applications, Materials Research Express. 2019;6:104007.

[36] Devaraju S, Krishnadevi K, Sriharshitha S, Alagar M, Design and Development of Environmentally Friendly Polybenzoxazine–Silica Hybrid from Renewable Bio-resource, Journal of Polymers and the Environment. 2019; 27:141-147.

[37] Krishnadevi K, Devaraju S, Sriharshitha S, Alagar M, Development and characterization of fully bio-based polybenzoxazine-silica hybrid composites for low-k and flame-retardant applications, Polymers for Advanced Technologies. 2019;30: 1856-1864.

[38] Lyu Y, Ishida H. Natural-sourced benzoxazine resins, homopolymers, blends and composites: A review of their synthesis, manufacturing and applications, Progress in Polymer Science, 2019, 99, 101168.

[39] Hariharan A, Srinivasan K, Murthy C, Alagar M. Cardanol Based Benzoxazine Blends and Bio-Silica Reinforced Composites: Thermal and Dielectric Properties, New Journal of Chemistry. 2018;42:4067-4080

[40] Sriharshitha S, Krishnadevi K, Devaraju S, Srinivasadesikan V, Lee SY, Eco-Friendly Sustainable Poly (benzoxazine-co-urethane) with Room-Temperature-Assisted Self-Healing Based on Supramolecular Interactions, ACS Omega. 2020;5:33178−33185.

[41] Wang C, Sun J, Liu X, Sudo A, Endo T. Synthesis and copolymerization of fully bio-based benzoxazines from guaiacol, furfurylamine and stearylamine. Green Chemistry. 2012;14:2799–2806.

[42] Calo E, Maffezzoli A, Mele G, Martina F, Mazzetto SE, Tarzia A. Synthesis of a novel cardanol-based benzoxazine monomer and environmentally sustainable production of polymers and bio-composites. Green Chemistry. 2007;9:754–759.

[43] Sharma P, Lochab B, Kumar D, Roy PK, Sustainable Bis-benzoxazines from Cardanol and PET-Derived Terephthalamides, ACS Sustainable Chemistry and Engineering. 2016;4: 1085–1093

[44] Froimowicz P, Arza CR, Han L, Ishida H. Smart, sustainable, and ecofriendly chemical design of fully bio-based thermally stable thermosets based on benzoxazine chemistry. ChemSusChem. 2016;9:1921−1928.

[45] Zhang K, Han M, Han L, Ishida H, Resveratrol-based tri-functional benzoxazines: Synthesis, characterization, polymerization, and thermal and flame retardant properties, European Polymer Journal. 2019;116: 526-533.

[46] Thirukumaran P, Parveen AS, Sarojadevi M. Synthesis and copolymerization of fully biobased benzoxazines from renewable resources, ACS Sustainable Chemistry and Engineering. 2014;2:2790-2801.

[47] Qian Z, Zheng Y, Li Q, Wang L, Fu, F. Amidation Way of Diphenolic Acid for Preparing Biopolybenzoxazine Resin with Outstanding Thermal Performance, Qian Z, Zheng Y, Li Q, Wang L, Fu, F. ACS Sustainable Chemistry and Engineering. 2021;9:12, 4668–4680.

[48] Zhang K, Han M, Liu Y, Formowicz P. Design and synthesis of bio-based high performance tri-oxazine benzoxazine resin via natural renewable resources, ACS Sustainable Chemistry and Engineering. 2019;7:9399-9407.

[49] Wen Z, Bonnaud L, Mincheva R, Dubois P, Raquez JM. Development of

Low-Viscosity and High-Performance Biobased Monobenzoxazine from Tyrosol and Furfurylamine. Materials. 2021;14:440

[50] Teng N, Yang S, Dai J, Wang S, Zhao J, Zhu J, Liu X. Making benzoxazine greener and stronger: Renewable resource, microwave irradiation, green solvent, and excellent thermal properties, ACS Sustainable Chemistry and Engineering. 2019;7:8715-8723.

[51] Periyasamy T, Asrafali KP, Muthusamy S, Kim SC. Replacing bisphenol-A with bisguaiachol-F to synthesize polybenzoxazines for a pollution-free environment, New Journal of Chemistry. 2016;40: 9313-9319.

[52] Li S, Yan S. Synthesis and characterization of novel biobased benzoxazines from cardbisphenol and the properties of their polymers. RSC Advances. 2015;5:61808–61814.

[53] Hariharan A, Ismail AAM, Latha G, Alagar M. Development of bio-based benzoxazines coated melamine foam for oil-water separation, Progress in Organic Coatings. 2021;53:106128.

[54] Dinesh Kumar G, Prabunathan P, Manoj M. et al. Fluorine Free Bio-Based Polybenzoxazine Coated Substrates for Oil-Water Separation and Anti-Icing Applications. Journal of Polymer and the Environment. 2020;28:2444–2456.

[55] Hariharan A, Prabunathan P, Kumaravel A, Manoj M, Alagar M. Bio-based polybenzoxazine composites for oil-water separation, sound absorption and corrosion resistance applications, Polymer Testing. 2020;86:106443.

[56] Periyasamy T, Asrafali SP, Shanmugam M, Kim SC, Development of sustainable and antimicrobial film based on polybenzoxazine and cellulose, International Journal of Biological Macromolecules. 2021;170:664-673.

[57] Yao H, Lu X, Xin Z, Li X, Chen C, Cao Y. Two novel eugenol-based difunctional benzoxazines: Synthesis and properties, Colloids and Surfaces A: Physicochemical and Engineering Aspects. 2021;616:126209.

[58] http://cashewindia.org/statistics

[59] Jaillet F, Darroman E, Ratsimihety A, Auvergne R, Boutevin B, Caillol S. New biobased epoxy materials from cardanol, European Journal of Lipid Science and Technology. 2014;116: 63-73.

[60] Liu Z, Chen J, Knothe G, Nie X, Jiang J. Synthesis of epoxidized cardanol and its antioxidative properties for vegetable oils and biodiesel, ACS Sustainable Chemistry and Engineering. 2016;4:901–906.

[61] Wang H, Zhou Q. Synthesis of Cardanol-Based Polyols via Thiol-ene/Thiol-epoxy Dual Click-Reactions and Thermosetting Polyurethanes Therefrom, ACS Sustainable Chemistry and Engineering. 2018;6:12088–12095

[62] Abbina S, Anilkumar P. Emergence of Sustainable Approaches for Functional Materials: Cashew Nut Shell Liquid and Other Relevant Crop-Based Renewable Resources. In: Anilkumar P. (eds) Cashew Nut Shell Liquid. Springer, Cham, 2017.

[63] Shukla S, Mahata A, Pathak B, Lochab B. Cardanol benzoxazines – interplay of oxazine functionality (mono to tetra) and properties, RSC Advances. 2015;5:78071–78080.

[64] Zhang C, Zhang Y, Zhou Q, Ling H, Gu Y. Processability and mechanical properties of bisbenzoxazine modified by the cardanol-based aromatic diamine benzoxazine, Journal of Polymer Engineering. 2014;34:561–568.

[65] Shukla S, Lochab B. Role of higher aromatic content in modulating

properties of cardanol based benzoxazines, Polymer. 2016;99: 684-694.

[66] Thirukumaran P, Sathiyamoorthi R, Parveen AS, Sarojadevi M. New benzoxazines from renewable resources for green composite applications, Polymer Composites. 2014;37:573-582.

[67] Lin C, Cai S, Leu T, Hwang T, Lee H. Synthesis and properties of flame-retardant benzoxazines by three approaches. Journal of Polymer Science Part A: Polymer Chemistry. 2016;44: 3454–3468.

[68] Spontón M, Ronda JC, Galià M, Cádiz V. Development of flame retardant phosphorus- and silicon-containing polybenzoxazines. Polymer Degradation and Stability. 2009;94:145–150.

[69] Ling H, Gu Y. Improving the flame retardancy of polybenzoxazines with a reactive phosphorus-containing compound. Journal of Macromolecular Science Part B Physics. 2011;50:2393–2404.

[70] Yan H, Sun C, Fang Z, Liu X, Zhu J, Wang H. Synthesis of an intrinsically flame retardant bio-based benzoxazine resin. Polymer. 2016;97:418–427.

[71] Amarnath N, Appavoo D, Lochab B. Eco-friendly halogen-free flame retardant cardanol polyphosphazene polybenzoxazine networks, ACS Sustainable Chemistry and Engineering. 2018;6:389–402.

[72] Appavoo D, Amarnath N, Lochab B, Cardanol and Eugenol Sourced Sustainable Non-halogen Flame Retardants for Enhanced Stability of Renewable Polybenzoxazines. Frontiers in Chemistry. 2020;8:711.

[73] Hariharan A, Prabunathan P, Manoj M, Alagar M. Studies on heterocyclic amines based cardanol-benzoxazine for oil-water separation, Polymer Engineering & Science. 2020; 60:1494-1506.

[74] Thirukumaran P, Parveen AS, Sarojadevi M. New Benzoxazines Containing Polyhedral Oligomeric Silsesquioxane from Eugenol, Guaiacol and Vanillin, New Journal of Chemistry. 2015;39:1691-1702.

[75] Devaraju S, Krishnadevi K, Naveena E, Alagar M. Eco-friendly fully bio-based polybenzoxazine-silica hybrid materials by sol–gel approach, Polymer Bulletin. 2020 (Article in Press).

[76] Amarnath N, Shukla S, Lochab B, Harvesting the Benefits of Inherent Reactive Functionalities in Fully Biosourced Isomeric Benzoxazines, ACS Sustainable Chemistry and Engineering. 2018;6:15151–15161.

[77] Guo W, Wang X, ChandraSekhar G, Wang J, Pan Y, Xing W, Song L, Hu Y. Cardanol derived benzoxazine in combination with boron-doped graphene toward simultaneously improved toughening and flame retardant epoxy composites, Composites Part A: Applied Science and Manufacturing. 2019;116:13-23.

[78] Yang R, Han M, Hao B, Zhang K, Biobased High-Performance Tri-Furan Functional Bis-Benzoxazine Resin Derived from Renewable Guaiacol, Furfural and Furfurylamine, European Polymer Journal. 2020;131:109706.

[79] Hao B, Han L, Liu Y, Zhang K, An Apigenin-Based Bio-Benzoxazine with Three Polymerizable Functionalities: Sustainable Synthesis, Thermal Latent Polymerization, and Excellent Thermal Properties of Its Thermosets, Polymer Chemistry. 2020;11:5800-5809.

[80] Zhang K, Liu Y, Han M, Froimowicz P, Smart and Sustainable Design of Latent Catalyst-Containing Benzoxazine Bio-Resins and Application

Studies, Green Chemistry. 2020;22: 1209-1219.

[81] Zhan Z, Yan H, Wang H, Cheng J, Ran S, Fang Z, Novel full bio-based phloroglucinol benzoxazine resin: Synthesis, curing reaction and thermal stability, Polymer. 2020;200:122534.

Chapter 4

Plant Uptake, Translocation and Metabolism of PBDEs in Plants

Daniel Dobslaw, Bertram Kuch, Christine Woiski and Martina Kiel

Abstract

Polybrominated diphenyl ethers (PBDEs) have been widely used as flame retardants in concentrations up to 30 w% of the total mass of the products. Worldwide consumption of technically relevant PBDE mixtures was about 7500 tons (penta-BDEs), 3790 tons (octa-BDEs) and 56,100 tons (deca-BDE) in 2001 and about 50–60% of this total volume was discharged into environment only by agricultural use of sewage sludges. The use of PBDEs was strictly regulated from 2004 onwards due to their high emission load and their effect as endocrine disrupters, neurotoxins, and fertility reducing agents. Nevertheless, soils worldwide are contaminated by gaseous and particle-bound transport of PBDEs. Therefore, the uptake of PBDEs from contaminated agricultural land via crops and the food chain is a major human exposure pathway. However, uptake and intrinsic transport behavior strongly depend on crop specifics and various soil parameters. The relevant exposure and transformation pathways, transport-relevant soil and plant characteristics and both root concentration factor (RCF) and transfer factor (TF) as derivable parameters are addressed and quantified in this chapter. Finally, based on available crop specific data a general statement about the transport behavior of PBDEs in twelve different crops according to relevant PBDE congeners is given.

Keywords: Plant uptake, translocation, root concentration factor, PBDE, shoot concentration factor, food industry, crops

1. Introduction

Polybrominated diphenyl ethers (PBDEs) have been used for decades as flame retardants in a wide variety of products. Notable among these are building insulations, upholstered furniture, electrical goods, vehicles and aircrafts, foams, textiles, electrical insulations, and a variety of technical plastics such as acrylonitrile-butadiene-styrene copolymers (ABS), high impact polystyrene (HIPS), polybutylene terephthalate (PBT), or plain paper (PAP), where PBDEs were used in concentrations of 5–30 percent by weight. Despite a spectrum of 209 PBDE congeners, only three formulations were of technical relevance, namely pentabromodiphenyl ether (penta-BDEs), octabromodiphenyl ether (octa-BDEs), and perbrominated diphenyl ether (deca-BDE, BDE-209). The global demand (EU demand) of these mixtures in 2001 was about 7500 tons (EU:150 tons), 3790 tons (EU: 610 tons), and 56,100 tons (EU: 7600 tons) [1]. The use of BDE-209 reached its peak in 2003–2006 with 30,000 tons (China), 9600 tons (EU), 5000–10,000 tons (Northern

America), and 1600 tons (Japan) [2]. Because of their endocrine-disrupting properties, neurotoxicity, and negative impacts on fertility as well as their high environmental persistence, the use of these mixtures was strictly regulated by the Stockholm Convention of 2001 and finally banned in 2004 (penta-BDEs, octa-BDEs) and severely restricted for BDE-209 in 2019 according to the lower toxicity in case of higher degree of bromination [3, 4].

As former high-volume chemicals PBDEs are ubiquitous in environment today, but are mainly detectable in soil and dust samples. E-waste sites and waste water sludge were identified as main sources with BDE-209 as dominant congener [5, 6]. Hence, BDE-209 levels of 6.3–12,194.6 ng g DM^{-1} (dry matter) at a ratio of 35–89.6% of the total PBDE where detected on e-waste recycling sites [7–9], while currently highest ∑PBDE levels of 8.70–18,451 ng g DM^{-1} were reported for soil at an industrial production site of plastic parts in electrical industry in Changzhou [10]. PBDE levels of 7240–10,469 ng g DM^{-1}, 180–370,000 ng g DM^{-1}, and 270–110,000 ng g DM^{-1} were also reported as currently highest levels in dust of industrial environment, house dust, and office dust samples in the UK, respectively [11]. In sludge samples of both municipal and industrial waste water treatment plants (WWTPs) in the USA, Turkey, and Hessian (Germany) levels of 85.5 ng g DM^{-1} up to 2.5 w% of ∑PBDE were reported [12–14]. The annual input of PBDEs into the environment in the USA was quantified as 47.9–60.1 tons, where 24.0–36.0 tons were located in sewage sludge [6].

As WWTP sludge is commonly used as fertilizer in agriculture, PBDE contamination is not restricted to hotspots like e-waste sites, and ubiquitous spread is further increased by gaseous and particulate-based transport of PBDEs. Consequently, soil samples were positively tested towards PBDE contaminations in grassland and forest soils of UK and Norway (∑PBDE: 65–12,000 pg. g DM^{-1}) [15], Western Austria (∑PBDE: 10.4–2744 pg. g DM^{-1}) [16], Germany (BDE-47: <27–505 pg. g DM^{-1}, BDE-209: <156–461 pg. g DM^{-1}) [17], and the Artic (∑12PBDEs ex BDE-209: 120 pg. g DM^{-1}, and ∑PBDE: 1.7–416 pg. g DM^{-1}) [17, 18].

Given the widespread distribution of PBDEs in soils, it can be assumed that they are absorbed by plants to a significant extent and are then introduced into humans via the food chain. The following sections are intended to show the factors influencing the plant uptake of PBDEs and their degradation products and to examine in detail the uptake and transport behavior of twelve selected crops. From these data, generally valid relationships for other crops will be derived using simple key parameters.

2. Physical characteristics and their effect on transport pathways and plant uptake

Because PBDEs show a wide range of molecular weight (328–959 g mol^{-1}), lipophilicity (log K_{OW} = 6–10), and volatility (log K_{OA} = 9–16) [19, 20], BDE congener specific transport and plant uptake mechanisms (soil-air-plant vs. soil–soil moisture-root-plant) are highly different and dependent on specific substance parameters (vapor pressure, Henry coefficient, air-plant partition coefficient, K_{OW} value, K_{OA} value), meteorological parameters (temperature, wind rate, precipitation, temporal rainfall distribution, deposition kinetics of gaseous and particulate BDEs), long range transport, plant specific characteristics (species, lipid content, carbohydrate content, fiber content, foliage morphology, non-lipid plant parts, rind consistency), and rhizosphere factors [18, 21–23]. Under aspects of transport, low brominated BDEs (Br_2-Br_3) are mainly and medium brominated BDEs (Br_4-Br_5), depending on the study, are minorly to dominantly distributed as gaseous compounds (BDE-15: 100%; BDE-28: 35–60%), while transmission and deposition of higher brominated congeners (Br_6-Br_{10}) are obligatorily characterized by adsorption of BDEs on a particulate phase [17, 22–25]. As a result of the particle-bound lower-range transport of the latter PBDEs, the PBDE pattern in soil and plant samples from densely populated regions and near hotspots shows high agreement with the present PBDE emission spectrum, while in sparsely populated regions the PBDE spectrum is dominated by low brominated congeners [7, 23]. Additionally, a significant concentration gradient of ∑PBDEs from both densely populated to sparsely populated regions and from emission sites to adjacent region can be observed [7, 26]. Hence, various studies around the world showed a wide range of PBDE concentrations in dust samples of 4.33–370,000 ng g DM^{-1} [11, 20, 27–31] with a clear domination of BDE-209 in the range of 69.2–99.6% [11, 20, 25].

Due to the high molar mass and the lipophilicity of high brominated BDEs, plant uptake by the soil–soil moisture-root-plant pathway is of low relevance and restricted to low and medium brominated BDEs (Br_2-Br_5) like BDE-47, BDE-99 and BDE-100 [21, 32], even though intrinsic transport of BDE-209 was reported by single studies [33–35], but disproved by Wu et al. [36], where plant availability of BDE-209 was quantified as 0.3–0.5% of the initial soil concentration and 99.5–99.7% of BDE-209 are solely adsorbed on the soil matrix and the outer side of the roots. Hence, atmospheric uptake of high brominated BDEs is the dominant pathway, even though BDE-209 reveals a low ratio of 0.1% of the atmospheric PBDE pattern [24, 37].

3. Human exposure to PBDE contaminations and uptake

As a result of the presence of gaseous and particulate PBDEs in air, the human PBDE uptake is dominated by inhalation, but the relevance of this pathway is strongly affected by atmospheric PBDE levels. Hites and Sjödin et al. reported concentrations of 5.27–301 pg. m^{-3} in ambient air and 0.06–67 ng m^{-3} at indoor air, but increased levels up to 312.1 ng BDE-209 m^{-3} at a Swedish e-waste recycling site [38, 39]. Average BDE-209 levels of 0.13 ng m^{-3} (gaseous) and 140 ng m^{-3} (particulate) were reported by Li et al. in 14 Chinese air samples and total BDE-209 uptake by inhalation was quantified as 3000 ng d^{-1} (respiration) and 69 ng d^{-1} (dust uptake), equivalent to 84% of the total daily uptake [40]. This finding was validated by multiple studies. As a consequence, 16% of daily PBDE uptake, and even higher ratios in case of lower ambient PBDE levels, are assigned to dietary uptake underlining the relevance as second dominant pathway.

4. Transformation and detoxification of PBDEs in plants

While PBDEs in the atmosphere are photolytically transformed by hydroxylation und subsequent transformation to lower brominated congeners or ring closure to the corresponding dibenzofurans [41, 42] and PBDEs in soil and sediments are mainly mineralized by stepwise debromination or detoxified by hydroxylation (OH-BDE) or methoxylation reactions (MeO-BDE) in the rhizosphere, strongly affected by the degree of bromination, concentration of oxygen, organic matter and microorganisms [43], intrinsic PBDEs in plants can be transformed by the same three transformation pathways. Exemplarily, transformation of BDE-28 and BDE-47 in maize was analyzed in detail by Wang et al. [44]. BDE-47 (Br_4) was dominantly converted to 6-MeO-BDE-47 (275 ng•g^{-1} DM) in the root phase, followed by 5-MeO-BDE-47 (40 ng•g^{-1} DM), ∑Br_2-BDEs (23 ng•g^{-1} DM), ∑Br_3-BDEs (20 ng•g^{-1}

DM), and small quantities of two unidentified hydroxylated BDEs (8 ng•g^{-1} DM) with continued decrease over time, similar to the transformation behavior against BDE-28 (Br_3). Similar results were observed in plants of pumpkins, rice, wheat and soybean for BDE-47 and BDE-99 with formation of 5-OH-BDE-47, 6-OH-BDE-47, 4'-OH-BDE-49, 4'-OH-BDE-42, 4-MeO-BDE-42, and BDE-28 in case of BDE-47 as parent congener, and with formation of 4-OH-BDE-99 and 4-MeO-BDE-99 in case of BDE-99 as soil contaminant [45, 46]. The total PBDE levels clearly dropped in all studies. In liver cells, transformation of both low and moderate brominated BDEs was shown by cytochrome P450 monooxygenases and glutathione-S-transferase, two enzyme complexes also found in plants, where they can potentially catalyze the same reactions. However, both enzyme sets were induced by BDE-209, but this congener was not converted [47]. In difference, various study showed comparable PBDE patterns both in soil and plant tissues at almost unchanged concentration levels over time, underlining negligible metabolism of PBDEs in plants [6]. In summary, PBDEs might be transformed in plants by debromination, hydroxylation and methoxylation reactions, but transformation behavior strongly depends on the plant species and the established microbial consortium in the rhizosphere [48].

5. Soil-root transport: RCF and TF value

5.1 RCF value of PBDEs

As a result of high bromination, high molecular weight, low mobility and high lipophilicity, BDE-209 as the dominant PBDE in soil is only marginally available for plants at levels of 0.3–0.5% of the initial concentration [36]. Despite the apparently low relevance, the soil – soil moisture – root uptake pathway is still of high relevance as tests with living and non-living roots of different plants showed 3.5–6 times higher BDE-209 levels in the living tissues [49]. Additional analysis of small-scale soil-based BDE gradients within the root plexus underlined the assumption of active BDE-209 uptake by plants [9] and was clearly proved by greenhouse experiments cultivating six different plant species in contaminated and non-contaminated soil in parallel [50]. In comparison to BDE-209 levels in contaminated soil, BDE-209 levels reached 5.2–10.4 ng g DM^{-1} and, therefore, less than 5%, i.e. more than 95% of BDE-209 contamination in plants could be attributed to root uptake and intrinsic plant transport. Both processes are coupled to plant transpiration elevating PBDE levels in shoots and leaves at dry weather conditions [51].

To increase comparability of PBDE uptake and intrinsic transport, both the root concentration factor (RCF) and the translocation factor (TF) were introduced in literature and correlated to the log K_{OW} value of PBDEs [9]. Both parameters can be correlated in a clearly negative way, i.e. higher RCF values were detected in case of lower brominated PBDEs and, therefore, compounds with lower log K_{OW} values than in case of higher log K_{OW} values [9]. In detail, plant specific RCFs of BDE-209 were up to ten times lower than the RCFs of BDE-28 [7, 9, 19]. This observation may be explained both by the lack of water solubility and thereby restricted root uptake with the soil moisture phase, and the strong adsorptive behavior of higher brominated PBDEs in soil. Hence, with exception of some plants like radish, green squash, and soft-stem bulrush, which are well adapted for phytoremediation processes, RCFs are clearly less than 1 for all PBDE congeners [52]. However, PBDE uptake is significantly affected by the presence of organic solubilizers in the soil like extracts of wheat straw or pig manure, where BDE-47 uptake by wheat as an example was elevated by a factor of 3.1 (wheat manure) and 1.9 (pig maure), respectively [53]. Hence, PBDE uptake increases with increasing surfactant activity, whereas decreases at weak surfactant activity and thereby increase of the organic content in soil [54]. Furthermore, PBDE uptake and RCFs as consequence are strongly affected by plant species specifics, physical and chemical soil properties, initial concentration levels of PBDEs, relevance of both gaseous and particulate atmospheric uptake, duration of the growth period, and organic soil content. Details will be described in Section 6.

In contrast, a positive correlation of both RCF and TF value was observed for maize in case of BDE-15, BDE-28, and BDE-47, which can be explained by an increased transpiration of the plants as these low-brominated congeners reveal higher water solubility [51].

5.2 TF value of PBDEs

The ratio of PBDE levels in shoots to the levels in roots is defined as translocation factor (TF). In contrast to the RCF, a general statement about the correlation of log K_{OW} and TF is not appropriate, since no clear positive or negative correlation occurs as the TF value depends on numerous parameters like plant species specifics, initial PBDE levels in soil, the lipid content of the shoots, plant age, distance of the plant issue from the root plexus, and the hardly quantifiable effect of the soil-air-plant exposure pathway. In principle, a negative correlation may be assumed as lower log K_{OW} values correlate with higher water solubility and, hence, higher intrinsic transport. As PBDE accumulate in the root area, stem and shoots show significantly lower contamination levels and relevance of atmospheric PBDE uptake significantly increases [51]. Nevertheless, the bioaccumulation and translocation behavior of PBDE in plants is not conclusively clarified and depends on numerous, partially insufficiently determined parameters.

6. Factors of PBDE plant uptake

Numerous studies focused on both physico-chemical and substance specific properties affecting plant uptake and biodegradation behavior of PBDE, where PBDE specifics (vapor pressure, K_{OW} value, air-water distribution K_{AW} value, air-plant distribution K_{AP} value), environmental parameters (temperature, wind rate, precipitation, temporal rain distribution, kinetics of both gaseous and particle-bound deposition), plant properties (species, lipid content, foliage morphology, ratio of non-lipid plant parts, rind thickness, contents of both sugar and fibers), as well as the presence of an microbial active rhizosphere were generally found to be highly germane. For bioavailability and thus biodegradability of PBDEs pH value and soil composition are of particular importance [18, 22, 23]. In detail, relevant parameters are:

6.1 Excretion of plant solubilizers

Easy metabolizable intermediates as amino acids, organic acids, sugars and exozenzymes are excreted by plants as detoxification strategy to improve microbial bioavailability and biodegradability of PBDE in the rhizosphere [33]. For example, hexose was excreted by *Kandelia obovate* to enhance microbial debromination of BDE-99 to Br_2-BDEs and Br_3-BDEs in soil [55].

6.2 Plant specifics

Plant specifics like plant morphology, wax layers of bay leaves and the lipid content of both leaf and roots strongly affect both atmospheric and soil-based uptake of PBDEs. Hence, accumulation of Br_3-BDEs to Br_{10}-BDEs in the wax layer of wheat was determined as 29–93% of the total plant uptake [25], while BDE-209 accumulation in six different plant species at initial soil levels of 4700 ng g DM^{-1} ranged 1822–10,933 ng $plant^{-1}$ with alfalfa showing lowest and maize showing highest levels [56].

6.3 Rhizosphere and mycorrhiza

The release of plant eluates into the mycorrhizal is part of the symbiosis between the plant and the mycorrhizal fungi promoting the plant's uptake of nutrients and the growth of microorganisms in the mycorrhizal area. Its secondary effect of enhanced microbial biodegradation and detoxification of PBDEs in the mycorrhizal area was shown several times in literature. As an example, increased depletion of 470 ng g DM^{-1} towards 2250 ng g DM^{-1} was observed for BDE-209 comparing rice plants with and without mycorrhizal fungi [57].

6.4 Specific root and leaf surface

A correlation between accumulation of PBDEs and high specific plant surface was clearly shown for BDE-209 in roots of radish, lettuce and taro [58, 59] and for Br_2-BDE to Br_{10}-BDE in both pine needles and eucalyptus leaves [60].

6.5 Lipid content

The lipid content of a plant, especially of the roots, shows strong effects on PBDE uptake and was successfully evaluated for various mosses, lichens, ryegrass, alfalfa, maize, radish, squash, and pumpkin. A positive correlation between lipid content and PBDE uptake (RCF), as well as a negative correlation in lipid content and intrinsic PBDE mobility (TF) was observed [19, 22, 61, 62].

6.6 Organic content of the soil

Similar to the lipid content an increase in organic content of soil evokes higher PBDE accumulation in the soil and, therefore, reduced PBDE plant uptake [20, 33, 63]. Exemplarily, PBDE uptake in carrots was reduced by 31.5–69.8% and soil-based biodegradation increased by 8.6–28.5% by addition of 1–4 w% of swine manure to the soil fraction [63]. Differentiation between TOC and DOC showed a clear improvement in adsorption of PBDE in the soil matrix at higher TOC levels, whereas no effect was observed in case of increased DOC levels [54].

6.7 Biochar

Since biochar is insoluble and thus is considered as a TOC increase of the soil, elevated accumulation of PBDE in the soil phase may occur [64].

6.8 Sewage sludge

As sewage sludge reveals a high TOC content, but also enhanced contamination levels with PBDE or their detoxification and degradation products, sewage sludge is a dominant PBDE exposure pathway. Hence, using contaminated sludges as agricultural fertilizers, increasing concentration levels and accumulation of PBDE in the soil phase and, finally, elevated PBDE plant uptake are observed. However, relevance of this effect strongly depends on the original PBDE contamination levels of the sludge. An overview of PBDE levels in different sewage sludges was previously presented [52]. Similar to both lipid content and TOC content, a negative correlation between BDE-209 uptake and organic content, implemented by sewage sludge out, was observed [37].

6.9 Compost and digestate

In difference to sewage sludge, the PBDE load of compost and digestates is rather low as confirmed by various studies due to the low contamination levels of the plant educts (leaves, green waste, fruit and food residues). For instance, a broadly based study of biocompost, green waste compost and digestates in Baden-Wuerttemberg showed comparable median concentrations of 13 ng g DM^{-1}, 5.4 ng g DM^{-1}, and 13.7 ng g DM^{-1} and confirmed to low relevance of these materials as PBDE emissions sources [65].

6.10 Soil humidity

As a result of the low water solubility of PBDE, high soil moisture effectively prevents evaporation of BDEs as well as plant uptake [64]. Correspondingly, a longer PBDE load may be expected at wet locations.

6.11 Plastic particles

Plastics may reveal partition coefficients of PBDEs of several orders of magnitude higher than those towards sewage sludge or soil at levels of 0.04–1.6 w% [66, 67]. In detail, accumulation of PBDEs in low density polyethylene films (LDPE), an adequate and commonly accepted model for diffusive transport phenomena across biomembranes, showed accumulation factors against the aqueous phase (K_{PEW}) of 10^5–10^7 in case of 23 BDE congeners covering the total spectrum of bromination [68]. Furthermore, a curvilinear correlation between the log K_{PEW} and log K_{OW} factor, very similar to the correlation of BCF and log K_{OW} shown in **Figure 1** was observed and explained by energy barriers for diffusive transport into the LDPE structure [68]. The statement of rising K_{PEW} levels at higher lipophilicity of BDE congeners was further confirmed [67]. However, K_{PEW} levels are strongly affected

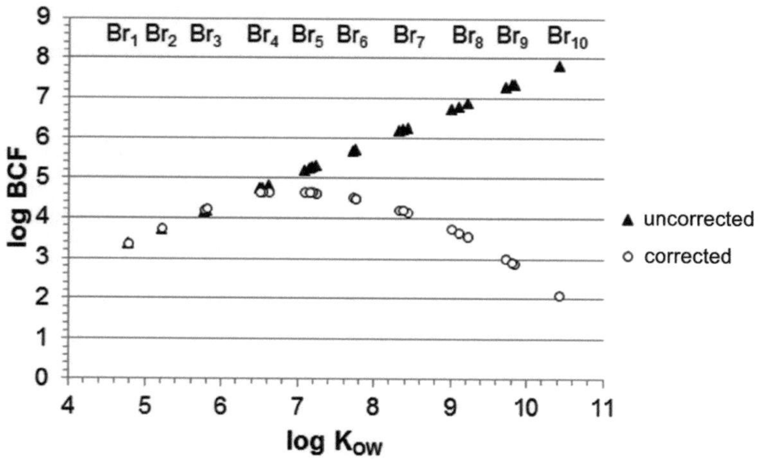

Figure 1.
Correlation of log K_{OW} and log BCF of 25 BDEs of high environmental relevance (−3, −7, −17, −28, −30, −47, −49, −66, −85, −99, −100, −123, −153, −154, −155, −183, −184, −191, −197, −201, −202, −206, −207, −208, −209) applying simple mathematical models with/without correction.

by the type of plastic and the temperature level during test conditions as adsorption is an endothermic process. While enhanced accumulation of BDE congeners was observed for polyethylene, polypropylene, polystyrene, and polyamide, low accumulation levels were observed for polyvinylchloride [66–68]. Hence, the hypothetical potential of injection of plastic particles into the soil as sink for PBDE and thereby soil remediation was positively investigated.

6.12 Other additives

Additional additives like graphene, TiO_2, Al_2O_3, Ag, and carbon nanotubes were considered as relevant for BDE-209 uptake in spinach, pumpkin, cucumber, corn and water spinach [36]. Indeed, an increased plant uptake was observed for all of these additives. Combinations of clay minerals like bentonite and oxidizing agents like sodium persulfate have also been positively tested for increased bioavailability of Br_3-Br_{10} – BDEs [69].

6.13 Solubilizers

The addition of surfactant-active additives elevates both mobility of PBDE in the soil matrix and plant uptake as previously described.

6.14 Macro- and trace elements

Apart from undisputed relevance of macro- and trace elements both in the development of the microflora in the rhizosphere and in plant growth, positive effects on the degradation of BDEs were observed in individual cases. Exemplarily, nitrate as additive caused intensified desorption and biodegradation of BDE-99, but might cause inhibition in case of high levels [70, 71].

6.15 Heavy metals

Presence of elevated concentrations of heavy metals shows ambivalent effects. Metals like Ni or Fe cause an enhanced uptake of PBDE, which was justified by chemical debromination of BDE-209 and enhanced mobilization, uptake and transport of Br_8- to Br_{10}-BDEs in the roots and shoots of the plants [64]. Enhanced plant uptake of low brominated BDEs like BDE-47 was also observed (24.76% vs. < 1.5%) [72]. In contrast, reduced BDE-209 uptake of up to 50% by pumpkins was reported after addition of 300 mg Cu kg DM^{-1} and microbial mineralization was negatively affected at more elevated levels [70, 73]. Similar effects were shown for lead, where BDE-209 uptake was reduced by a factor of 2.9–3.7 by tall fescue at levels up to 1950 mg Pb kg DM^{-1} [74]. Heavy metal induced effects on PBDE plant uptake are also plant specifics as shown for cadmium, where levels of up to 14,800 ng g DM-1 had no effect on BDE-209 uptake in case of black nightshade, but lifting effects in case of amaranth [75, 76].

In summary, presence of essential heavy metals like iron or copper at adequate concentrations might have a positive effect on PBDE degradation, while non-essential heavy metals at non-toxic levels reveal no effect.

7. Predictive mathematical models

Due to the broad spectrum of food plants, attempts were made to develop simple but sensitive models to predict PBDE plant uptake based on simple chemical conditions and input variables as distribution equilibria, lipid content, organic matter, and initial PBDE soil-water concentration to achieve predictive statements about RCF, SCF (shoot concentration factor) or TF.

While these models provide comparatively good correlations for the RCF, they commonly fail in prediction of the TF, because this value is strongly influenced by intrinsic and atmospheric transport of BDEs in addition to the plant specific uptake of PBDE in the root plexus. Therefore, deviations in the range of two decades can be observed comparing model and real situation [77, 78]. Even after restriction of models to single pollutant situations instead of congener mixtures with variable concentration levels, and after focusing on single and simplified plants like lettuce, where differentiations between shoot and fruit or over the height of the shoot are not applicable, deviations of 25.3–58.2% of the model compared to real situation were reported for the insecticide chlorpyrifor [79].

Another highly relevant error is caused by incorrect analysis of intrinsic PBDE levels in roots in contrast to adsorptive fractions at the outer root surface affecting quality of environmental data. Hence, Briggs et al. [80] showed a significant decrease in BCF levels and thus RCF values of PBDE starting at a log K_{OW} value of approx. 6.5 (corresponds to a log BCF value of approx. 4.6) after elimination of externally adsorbed congeners (see **Figure 1**). This chart corresponds to Bintein's bilinear model [81], which was confirmed by Meylan et al. [82] for 610 non-ionic pollutants. This negative correlation at high log K_{OW} values and thus high lipophilicity bases on three restrictions of lipophilic compounds as follows:

- **Equilibrium kinetics**: The higher the lipophilicity of a pollutant, the longer it takes to achieve equilibrium state between two phases or compartments. The life span of annual crops might be too short to establish equilibria between soil and root or root and shoot [83].

- **Solubility:** Water solubility decreases with increasing lipophilicity and strongly lipophilic substances are primarily adsorb onto particles or surfaces. However, for absorptive root uptake of pollutants, a phase transition from soil to the liquid phase as well as from the liquid phase to the intrinsic root is required without adsorptive elimination at the tissue [80, 83]

- **Membrane permeability and cellular transport mechanisms**: The membrane-based cellular uptake of pollutants takes place by passive permeation [80]. The membrane permeability and thus bioavailability of contaminants is concisely described by Lipinski's 'Law of 5', stating out low absorption or membrane permeability at log K_{OW} values higher than 5, molecular weight higher than 500, more than 5 hydrogen bond donors and more than 10 ($= 2 \cdot 5$) hydrogen bond acceptors. The former two requirements are fulfilled even in case of Br_4- to Br_5 – BDEs. By means of known transport mechanisms into the cell, PBDE plant uptake may be affected by co-transport phenomena of biomolecules like amino acids.

8. RCF and TF values of specific crops

Following the extensive literature evaluation by Dobslaw et al. [52] twelve crops with the highest documented data density were selected and the occurring RCF and TF values for BDE-47 and BDE-209 were compared. The highest data density regarding the transition of PBDEs from soil to root or from root to plant was available for BDE-209 followed by BDE-47. In contrast to lower brominated

BDEs the concentrations of BDE-209 found in literature range over several decades, which facilitates an evaluation of the literature data concerning transition rates. In particular, the following species were selected: Rice (*Oryza sativa* L.), maize (*Zea mays* L.), prince's-feather (*Amaranthus hypochondriacus* L.), the group of bok choy, Chinese cabbage, broccoli, and filder cabbage (*Brassica* sp.), lettuce (*Lactuca sativa sp.), spinach* (*Spinacia oleracea* L.), sweet potato vine (*Ipomoea batatas* L.), radish (*Raphanus sativus* L.), carrot (*Daucus carota* L.), taro root (*Colocasia esculenta* L. Schott), pea (*Pisum sativum* L.), and pumpkin (*Cucurbita pepo ssp. pepo*).

Figure 2 shows the concentrations of BDE-209 in the roots of these crops as a function of the corresponding soil concentrations (RCF), indicating the dependency of BDE-209 uptake on the soil concentration with a ratio of about 1:10 (c_{root}:c_{soil}). Comparing shoots and corresponding soil a similar dependency on soil concentration with a ratio of about 1:18 (c_{root}:c_{soil}) was found (**Figure 3**). This representation was chosen because of the higher data density being available. However,

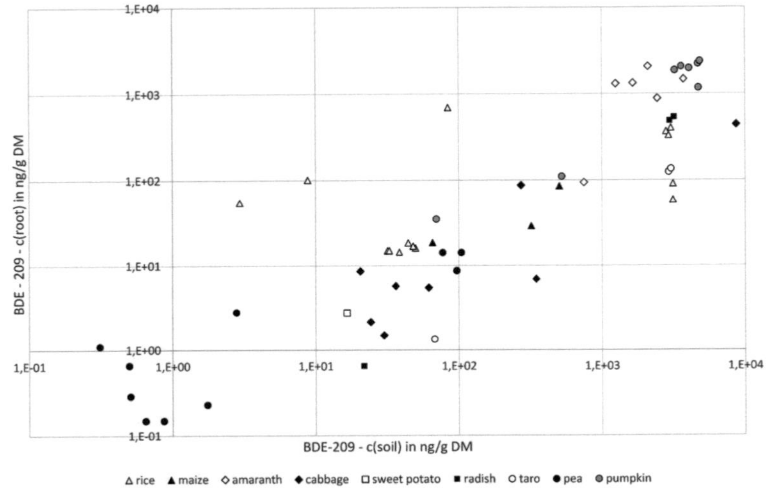

Figure 2.
BDE-209 – Root concentration factor (RCF), concentration in root as function of corresponding soil concentration.

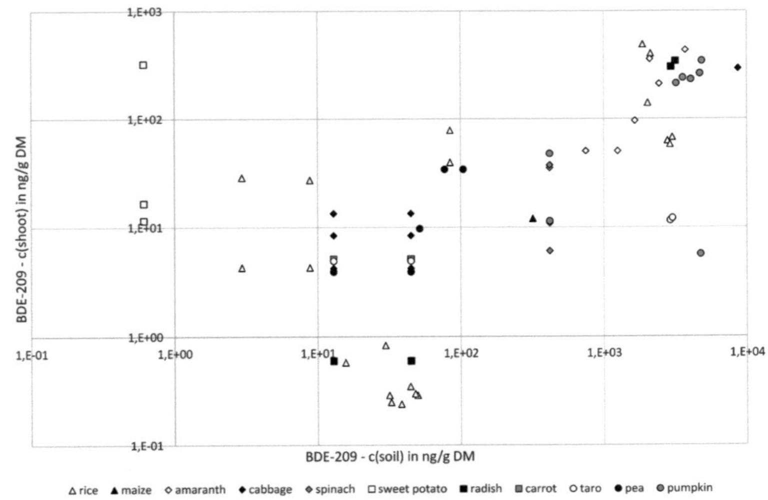

Figure 3.
BDE-209 – Concentration in shoot as function of corresponding soil concentration.

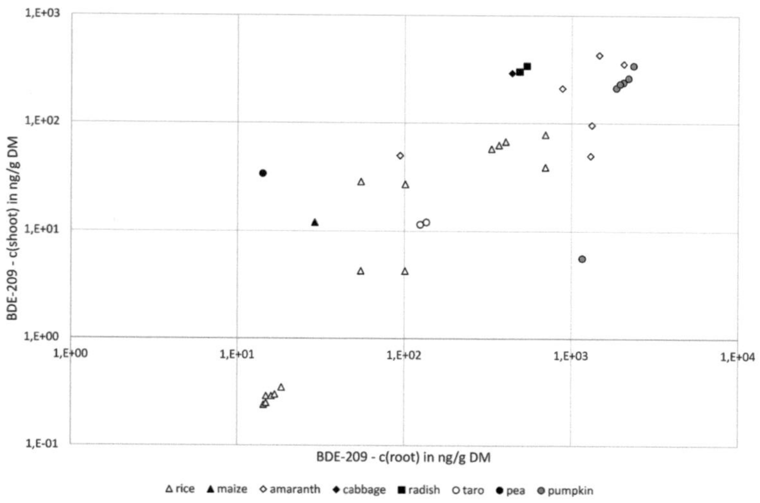

Figure 4.
BDE-209 – Translocation factor (TF), concentration in shoot as function of corresponding root concentration.

a TF could be derived from the comparison of the existing data pairs, which points to the dependence of the shoot concentrations on the concentrations in the root (**Figure 4**) with a ratio of about 1:9 (c_{shoot}:c_{root}).

The extent of transition of the tetrabrominated BDE-47 from soil to root and shoot was also determined by the concentration of the flame retardant in the soil (**Figures 5** and **6**). For BDE-47, despite its lower logK_{OW} in comparison to BDE-209, the transition rates are higher with about 1.1:1 (c_{root}:c_{soil}) and 1:6 (c_{shoot}:c_{soil}). Too few corresponding data pairs were available for the representation of the TF. A graphical summary of the results for the congeners 47 and 209 is given in **Figure 7**. The results are consistent with the prediction models described in Section 7 (see also **Figure 1**). Due to its high log K_{OW}, BDE-209 is expected to have a lower uptake compared to BDE-47. With the models mentioned above, it is possible to adequately describe the uptake behavior of lipophilic PBDEs. Of particular interest is the trend that the uptake of the substances is primarily determined by the

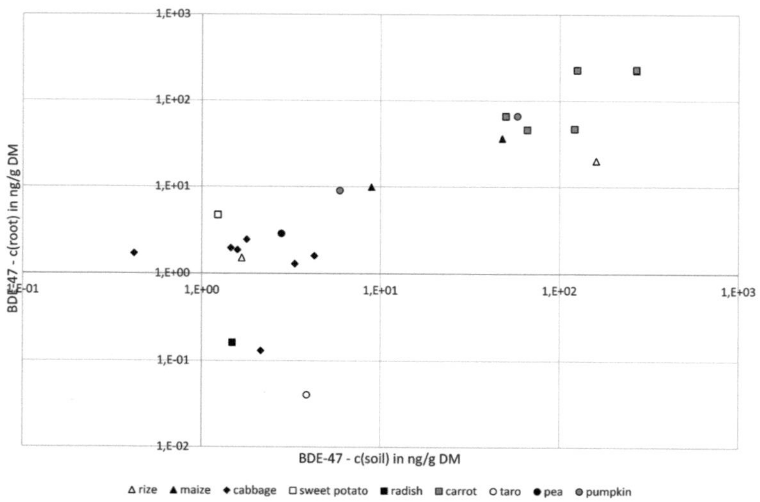

Figure 5.
BDE-47 – Root concentration factor (RCF), concentration in root as function of corresponding soil concentration.

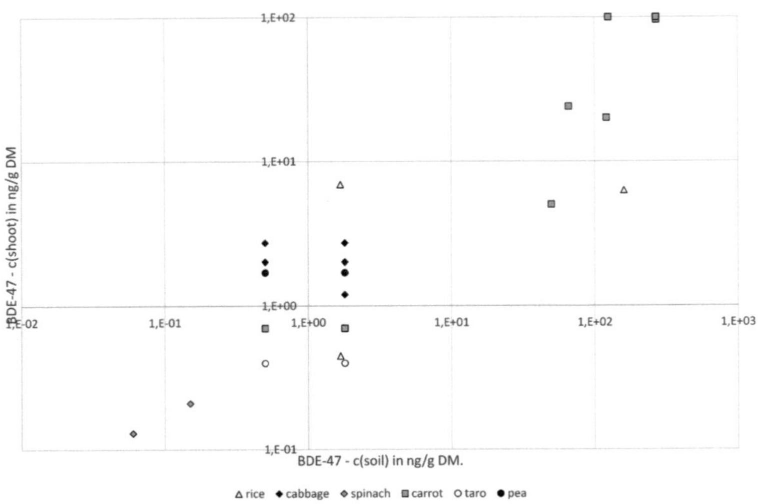

Figure 6.
BDE-47 – Concentration in shoot as function of corresponding soil concentration.

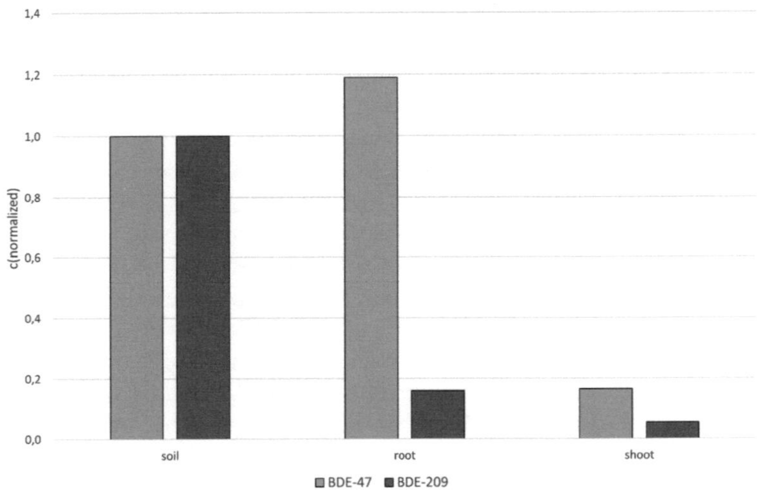

Figure 7.
BDE-47 and BDE-209 – Comparison of concentrations in soil, root and shoot (normalized, c(soil) =1).

concentration of the compounds in soil. A dependency of the uptake on the crop species might have been expected, but cannot be proven by the present data set.

9. Conclusion

PBDEs were widely used as flame retardants. Due to their effects as endocrine disruptors, neurotoxins, and on reproductive capacity as well, knowledge of terrestrial accumulation behavior and, in particular, their uptake by plants for food production is highly relevant. Uptake can occur viasoil-air-plant pathway as well as soil–soil water-root-plant pathway. Transport and plant uptake behavior strongly depend on physical and chemical properties of the BDEs, environmental factors, large-scale atmospheric transport processes, plant properties as well as terrestrial rhizospheres. During both atmospheric and terrestrial transport PBDEs are subject to UV-induced (atmospheric) or microbial induced (terrestrial) transformation and degradation processes like debromination, hydroxylation, methoxylation and ring closure to dibenzofurans. As PBDEs reveal high lipophilicity they tend to adsorption on lipophilic soil matrices and thereby show low uptake via soil–soil water-root-plant pathway and subsequent intrinsic transport. Hence, uptake and intrinsic transport are only expected for low brominated BDEs. Therefore, declining concentrations of PBDEs could be detected from soil via roots to shoots and final fruits, i.e. RCF and TF show negative correlation with their log K_{OW} values. Consistent with this statement, 84% of the human PBDE intake are attributed to respiration and inhalation of dust, while only 16% were correlated with dietary uptake. The actual exposure of vegetarian foods to PBDEs depends on the following parameters:

- Both microbial degradation and plant uptake of PBDE are elevated by release of easily biodegradable plant extracts like amino acids, organic acids, sugars and exoenzymes as commonly observed in symbiosis with rhizobia.

- The presence of rhizobia enhances microbial degradation and plant uptake of PBDE.

- Both atmospheric and terrestrial PBDE uptakes are strongly affected by the morphology of the plant (specific surface of leaves and roots) and the increasing lipid content of plant tissues accelerating PBDE uptake.

- PBDE immobilization and accumulation in soil is promoted by increasing TOC levels caused by implementation of compost, sewage sludge, digestates, or biochar.

- In contrast, increasing levels of DOC show no effect on plant uptake of PBDEs as long as there is no solubilizing effect by surfactants.

- As sewage sludge shows PBDE contamination up to 2.5 w%, it is an important emission source of PBDEs affecting soil contamination. In contrast, PBDE loads of compost and digestates are very low and commonly pollution effects are negligible.

- A decrease in evaporation losses and an enhanced immobilization tendency of PBDE can be observed in case of raising soil moisture.

- Solubilizers, ionic additives and nanoscale organic substances act as mobilizing agents increasing mobilization and plant uptake of PBDEs.

- Macroelements as nitrate favor terrestrial PBDE degradation through its function as alternative electron acceptor. Hence, plant's load is reduced.

- The presence of trace elements supports microbial transformation of PBDE in soil and therefore causes lower contamination of plants. In contrast, heavy metals seem to enforce PBDE uptake by plants by inhibition of terrestrial biodegradation processes.

- Current mathematical models allow high quality in prediction of the RCF value with a minimum of input parameters. In opposite, the prediction of SCF and TF values is not suitable due to the insufficient coverage of plant-specific parameters.

- According to Lipinski's 'Law of 5' Br_4- and Br_5-BDE congeners reveal the highest RCF levels. Higher polarity is required to gain high TF factors. Within the same isomers, even small differences in lipophilicity significantly change these values.

- Available PBDE data only show a linear correlation between soil concentration and plant uptake. Plant specific uptake effects cannot be observed.

10. Outlook

The poor biodegradability of PBDEs means that both the accumulation of PBDEs in the food chain and inhalation exposure will continue to be highly relevant issues in the next few decades. Problems comparable to those described for PBDEs are also expected for alternative brominated flame retardants such as hexabromobenzene, pentabromotoluene, 1,2-bis-(2,4,6-tribromophenoxy)ethane, or decabromodiphenylethane due to the likewise high degree of bromination, high persistence and thus a high bioaccumulation potential [20, 84]. A ubiquitous presence of these compounds was already proven [19].

Especially remarkable is the partial lack of metadata in existing publications, which makes an comprehensive evaluation, as it was tried in this publication, difficult. This concerns, for example, more detailed information about the soil composition, the content of organic carbon, lipid, and water of roots and shoots. Thinning of data, such as the aggregation of single congener data to sums (e.g. ∑PBDE) without further information about the composition makes a valuable data set useless.

Acknowledgements

We would like to thank Lukas Lesmeister (DVGW, Karlsruhe), Melanie Mechler and Jörn Breuer (Center of Agricultural Technology Augustenberg, Karlsruhe) and Prof. Karl-Heinrich Engesser (ISWA, Stuttgart) for their continuous support. This work was supported by the German Federal Environment Agency at grant number FKZ 3718 74 210 0; AZ: 91 007 – 2/82.

Conflict of interest

The authors declare that they have no conflict of interest.

Author details

Daniel Dobslaw*, Bertram Kuch, Christine Woiski and Martina Kiel
Institute of Sanitary Engineering, Water Quality and Solid Waste Management, University of Stuttgart, Stuttgart, Germany

*Address all correspondence to: daniel.dobslaw@iswa.uni-stuttgart.de

References

[1] BSEF Bromine Science and Environmental Forum. Major Brominated Flame Retardants Volume Estimates - Total Market Demand by Region in 2001

[2] ECHA European Chemicals Agency. Background document to the Opinion on the Annex XV dossier proposing restrictions on Bis(pentabromophenyl) ether: ECHA/RAC/RES-O-0000006155-77-01/D ECHA/SEAC/RES-O-0000006155-77-03/F. Helsinki; 2015

[3] Sun Y, Wang C, Xu X, Ruan H. Responses of plants to polybrominated diphenyl ethers (PBDEs) induced phytotoxicity: A hierarchical meta-analysis. Chemosphere. 2020;**240**:124865. DOI: 10.1016/j.chemosphere.2019.124865

[4] European Commission EC. EC Regulation 217/227 amending Annex XVII to Regulation (EC) No 1907/2006 of the European Parliament and of the Council concerning the Registration, Evaluation, Authorisation and Restriction of Chemicals (REACH) as regards bis(pentabromophenyl)ether.

[5] Stone R. China. Confronting a toxic blowback from the electronics trade. Science (New York, N.Y.). 2009;**325**:1055. DOI: 10.1126/science.325_1055

[6] Venkatesan AK, Halden RU. Brominated flame retardants in U.S. biosolids from the EPA national sewage sludge survey and chemical persistence in outdoor soil mesocosms. Water research. 2014;**55**:133-142. DOI: 10.1016/j.watres.2014.02.021

[7] Han Z-X, Wang N, Zhang H-L, Zhao Y-X. Bioaccumulation of PBDEs and PCBs in a small food chain at electronic waste recycling sites. Environmental Forensics. 2017;**18**:44-49. DOI: 10.1080/15275922.2016.1263900

[8] Leung AOW, Luksemburg WJ, Wong AS, Wong MH. Spatial distribution of polybrominated diphenyl ethers and polychlorinated dibenzo-p-dioxins and dibenzofurans in soil and combusted residue at Guiyu, an electronic waste recycling site in southeast China. Environmental Science & Technology. 2007;**41**:2730-2737. DOI: 10.1021/es0625935

[9] Zhang Y, Luo X-J, Mo L, Wu J-P, Mai B-X, Peng Y-H. Bioaccumulation and translocation of polyhalogenated compounds in rice (Oryza sativa L.) planted in paddy soil collected from an electronic waste recycling site, South China. Chemosphere. 2015;**137**:25-32. DOI: 10.1016/j.chemosphere.2015.04.029

[10] Xu J, Qian W, Li J, Zhang X, He J, Kong D. Polybrominated diphenyl ethers (PBDEs) in soil and dust from plastic production and surrounding areas in eastern of China. Environmental geochemistry and health. 2019;**41**:2315-2327. DOI: 10.1007/s10653-019-00247-0

[11] Tao F, Abdallah MA-E, Harrad S. Emerging and Legacy Flame Retardants in UK Indoor Air and Dust: Evidence for Replacement of PBDEs by Emerging Flame Retardants? Environmental Science & Technology. 2016;**50**:13052-13061. DOI: 10.1021/acs.est.6b02816

[12] Hale RC, La Guardia MJ, Harvey E, Chen D, Mainor TM, Luellen DR, Hundal LS. Polybrominated diphenyl ethers in U.S. sewage sludges and biosolids: temporal and geographical trends and uptake by corn following land application. Environmental Science & Technology. 2012;**46**:2055-2063. DOI: 10.1021/es203149g

[13] Leisewitz A, Fengler S, Seel P. Orientierende Messungen gefährlicher Stoffe: Landesweite Untersuchung auf organische Spurenverunreinigungen in hessischen Fließgewässern, Abwässern und Klärschlämmmen.: Zusammenfassender Abschlussbericht 1991 - 2003. Wiesbaden: Hessisches Landesamt für Umwelt und Geologie (HLUG); 2003

[14] Demirtepe H, Imamoglu I. Levels of polybrominated diphenyl ethers and hexabromocyclododecane in treatment plant sludge: Implications on sludge management. Chemosphere. 2019;**221**:606-615. DOI: 10.1016/j.chemosphere.2019.01.060

[15] Hassanin A, Breivik K, Meijer SN, Steinnes E, Thomas GO, Jones KC. PBDEs in European Background Soils: Levels and Factors Controlling Their Distribution. Environmental Science & Technology. 2004;**38**:738-745. DOI: 10.1021/es035008y

[16] Freudenschuß A, Obersteiner Erik, Uhl Maria. Organische Schadstoffe in Grünlandböden: TEIL 3 – Endbericht. Wien: Umweltbundesamt GmbH; 2010

[17] Dreyer A, Neugebauer F, Rüdel H, Klein R, Lohmann N, Rauert C, Koschorreck J. Halogenated flame retardants in tree samples applied as bioindicators for atmospheric pollution. Chemosphere. 2018;**208**:233-240. DOI: 10.1016/j.chemosphere.2018.05.033

[18] Zhu C, Li Y, Wang P, Chen Z, Ren D, Ssebugere P, Zhang Q, Jiang G. Polychlorinated biphenyls (PCBs) and polybrominated biphenyl ethers (PBDEs) in environmental samples from Ny-Ålesund and London Island, Svalbard, the Arctic. Chemosphere. 2015;**126**:40-46. DOI: 10.1016/j.chemosphere.2015.01.043

[19] She Y-Z, Wu J-P, Zhang Y, Peng Y, Mo L, Luo X-J, Mai B-X. Bioaccumulation of polybrominated diphenyl ethers and several alternative halogenated flame retardants in a small herbivorous food chain. Environmental Pollution. 2013;**174**:164-170. DOI: 10.1016/j.envpol.2012.11.024

[20] Zhu H, Sun H, Yao Y, Gan Z, Wang Y, Kannan K. Legacy and alternative brominated flame retardants in outdoor dust and pine needles in mainland China: Spatial trends, dust-plant partitioning and human exposure. Environmental pollution (Barking, Essex: 1987). 2018;243:758-765. DOI: 10.1016/j.envpol.2018.08.097

[21] Klinčić D, Dvoršćak M, Jagić K, Mendaš G, Herceg Romanić S. Levels and distribution of polybrominated diphenyl ethers in humans and environmental compartments: a comprehensive review of the last five years of research. Environmental Science and Pollution Research. 2020;**27**:5744-5758. DOI: 10.1007/s11356-020-07598-7

[22] Yogui GT, Sericano JL, Montone RC. Accumulation of semivolatile organic compounds in Antarctic vegetation: a case study of polybrominated diphenyl ethers. The Science of the total environment. 2011;**409**:3902-3908. DOI: 10.1016/j.scitotenv.2011.06.010

[23] Zhao Y-X, Qin X-F, Li Y, Liu P-Y, Tian M, Yan S-S, Qin Z-F, Xu X-B, Yang Y-J. Diffusion of polybrominated diphenyl ether (PBDE) from an e-waste recycling area to the surrounding regions in Southeast China. Chemosphere. 2009;**76**:1470-1476. DOI: 10.1016/j.chemosphere.2009.07.023

[24] Gao M, Wang G, Lin B, Tariq M, Liu K, Zhang W. Study on arbor leaf and ring as a potential biological indicator for atmospheric polybrominated diphenyl ethers (PBDEs) distribution at e-wastes recycling sites. International Journal of Environmental Science and

Technology. 2019;**16**:8639-8652. DOI: 10.1007/s13762-019-02428-x

[25] Zhu H, Wang F, Li B, Yao Y, Wang L, Sun H. Accumulation and translocation of polybrominated diphenyl ethers into plant under multiple exposure scenarios. Environment international. 2020;**143**:105947. DOI: 10.1016/j.envint.2020.105947

[26] Graziani NS, Tames MF, Mateos AC, Silva JA, Ramos S, Homem V, Ratola N, Carreras H. Estimation of urban POP and emerging SVOC levels employing Ligustrum lucidum leaves. Atmospheric Pollution Research. 2019;**10**:1524-1530. DOI: 10.1016/j.apr.2019.04.010

[27] Wang J, Ma Y-J, Chen S-J, Tian M, Luo X-J, Mai B-X. Brominated flame retardants in house dust from e-waste recycling and urban areas in South China: implications on human exposure. Environment international. 2010;**36**:535-541. DOI: 10.1016/j.envint.2010.04.005

[28] Ma J, Addink R, Yun S, Cheng J, Wang W, Kannan K. Polybrominated dibenzo-p-dioxins/dibenzofurans and polybrominated diphenyl ethers in soil, vegetation, workshop-floor dust, and electronic shredder residue from an electronic waste recycling facility and in soils from a chemical industrial complex in eastern China. Environmental Science & Technology. 2009;**43**:7350-7356. DOI: 10.1021/es901713u

[29] Muenhor D, Harrad S, Ali N, Covaci A. Brominated flame retardants (BFRs) in air and dust from electronic waste storage facilities in Thailand. Environment international. 2010;**36**:690-698. DOI: 10.1016/j.envint.2010.05.002

[30] Schreder ED, La Guardia MJ. Flame retardant transfers from U.S. households (dust and laundry wastewater) to the aquatic environment. Environmental Science & Technology. 2014;**48**:11575-11583. DOI: 10.1021/es502227h

[31] Harrad S, Goosey E, Desborough J, Abdallah MA-E, Roosens L, Covaci A. Dust from U.K. primary school classrooms and daycare centers: the significance of dust as a pathway of exposure of young U.K. children to brominated flame retardants and polychlorinated biphenyls. Environmental Science & Technology. 2010;**44**:4198-4202. DOI: 10.1021/es100750s

[32] Mueller KE, Mueller-Spitz SR, Henry HF, Vonderheide AP, Soman RS, Kinkle BK, Shann JR. Fate of Pentabrominated Diphenyl Ethers in Soil: Abiotic Sorption, Plant Uptake, and the Impact of Interspecific Plant Interactions. Environmental Science & Technology. 2006;**40**:6662-6667. DOI: 10.1021/es060776l

[33] Zhao L, Jiang J, Chen C, Zhan S, Yang J, Yang S. Efficiency and mechanism of the phytoremediation of decabromodiphenyl ether-contaminated sediments by aquatic macrophyte Scirpus validus. Environmental science and pollution research international. 2017;**24**:12949-12962. DOI: 10.1007/s11356-017-8900-1

[34] Vrkoslavová J, Demnerová K, Macková M, Zemanová T, Macek T, Hajslová J, Pulkrabová J, Hrádková P, Stiborová H. Absorption and translocation of polybrominated diphenyl ethers (PBDEs) by plants from contaminated sewage sludge. Chemosphere. 2010;**81**:381-386. DOI: 10.1016/j.chemosphere.2010.07.010

[35] Chow KL, Man YB, Tam NFY, Liang Y, Wong MH. Uptake and transport mechanisms of decabromodiphenyl ether (BDE-209) by rice (Oryza sativa). Chemosphere. 2015;**119**:1262-1267. DOI: 10.1016/j.chemosphere.2014.10.016

[36] Wu X, Wang W, Zhu L. Enhanced organic contaminants accumulation in crops: Mechanisms, interactions with engineered nanomaterials in soil. Environmental Pollution. 2018;**240**:51-59. DOI: 10.1016/j.envpol.2018.04.072

[37] Li H, Qu R, Yan L, Guo W, Ma Y. Field study on the uptake and translocation of PBDEs by wheat (*Triticum aestivum* L.) in soils amended with sewage sludge. Chemosphere. 2015;**123**:87-92. DOI: 10.1016/j.chemosphere.2014.12.045

[38] Hites RA. Polybrominated diphenyl ethers in the environment and in people: a meta-analysis of concentrations. Environmental Science & Technology. 2004;**38**:945-956. DOI: 10.1021/es035082g

[39] Sjödin A, Carlsson H, Thuresson K, Sjölin S, Bergman A, Ostman C. Flame retardants in indoor air at an electronics recycling plant and at other work environments. Environmental Science & Technology. 2001;**35**:448-454. DOI: 10.1021/es000077n

[40] Li P, Wu H, Li Q, Jin J, Wang Y. Brominated flame retardants in food and environmental samples from a production area in China: concentrations and human exposure assessment. Environmental monitoring and assessment. 2015;**187**:719. DOI: 10.1007/s10661-015-4947-y

[41] Ueno D, Darling C, Alaee M, Pacepavicius G, Teixeira C, Campbell L, Letcher RJ, Bergman A, Marsh G, Muir D. Hydroxylated polybrominated diphenyl ethers (OH-PBDEs) in the abiotic environment: surface water and precipitation from Ontario, Canada. Environmental Science & Technology. 2008;**42**:1657-1664. DOI: 10.1021/es7021279

[42] Kuch B, Schneider C, Rupp S, Recke Rvd, Bopp K, Metzger JW. Polybromierte Diphenylether und Tetrabrombisphenol A: Untersuchungen zum Abbau und Metabolismus, Bestimmung in Nahrungsmitteln: Abschlussbericht BWPLUS. Karlsruhe: Landesanstalt für Umwelt Baden-Württemberg (LUBW); 2005

[43] Zhu H, Wang Y, Tam NFY. Microcosm study on fate of polybrominated diphenyl ethers (PBDEs) in contaminated mangrove sediment. Journal of hazardous materials. 2014;**265**:61-68. DOI: 10.1016/j.jhazmat.2013.11.046

[44] Wang S, Zhang S, Huang H, Lu A, Ping H. Debrominated, hydroxylated and methoxylated metabolism in maize (Zea mays L.) exposed to lesser polybrominated diphenyl ethers (PBDEs). Chemosphere. 2012;**89**:1295-1301. DOI: 10.1016/j.chemosphere.2012.05.026

[45] Pan L, Sun J, Wu X, Wei Z, Zhu L. Transformation of hydroxylated and methoxylated 2,2′,4,4′,5-brominated diphenyl ether (BDE-99) in plants. Journal of Environmental Sciences. 2016;**49**:197-202. DOI: 10.1016/j.jes.2016.06.017

[46] Sun J, Liu J, Yu M, Wang C, Sun Y, Zhang A, Wang T, Lei Z, Jiang G. In vivo metabolism of 2,2′,4,4′-tetrabromodiphenyl ether (BDE-47) in young whole pumpkin plant. Environmental Science & Technology. 2013;**47**:3701-3707. DOI: 10.1021/es4003263

[47] Stapleton HM, Kelly SM, Pei R, Letcher RJ, Gunsch C. Metabolism of polybrominated diphenyl ethers (PBDEs) by human hepatocytes in vitro. Environmental health perspectives. 2009;**117**:197-202. DOI: 10.1289/ehp.11807

[48] Chen J, Zhou HC, Wang C, Zhu CQ, Tam NF-Y. Short-term enhancement effect of nitrogen addition on microbial degradation and plant uptake of

polybrominated diphenyl ethers (PBDEs) in contaminated mangrove soil. Journal of hazardous materials. 2015;**300**:84-92. DOI: 10.1016/j.jhazmat.2015.06.053

[49] Chow KL, Man YB, Tam NFY, Liang Y, Wong MH. Removal of decabromodiphenyl ether (BDE-209) using a combined system involving TiO2 photocatalysis and wetland plants. Journal of hazardous materials. 2017;**322**:263-269. DOI: 10.1016/j.jhazmat.2016.05.097

[50] Jiang J, Zhao H, Sun S, Wang Y, Liu S, Xie Q, Li X. Occurrence and profiles of halogenated phenols, polybrominated diphenyl ethers and hydroxylated polybrominated diphenyl ethers in the effluents of waste water treatment plants around Huang-Bo Sea, North China. The Science of the total environment. 2018;**622-623**:1-7. DOI: 10.1016/j.scitotenv.2017.11.323

[51] Zhao M, Zhang S, Wang S, Huang H. Uptake, translocation, and debromination of polybrominated diphenyl ethers in maize. Journal of Environmental Sciences. 2012;**24**:402-409. DOI: 10.1016/S1001-0742(11)60748-1

[52] Dobslaw D, Woiski C, Kiel M, Kuch B, Breuer J. Plant uptake, translocation and metabolism of PBDEs in plants of food and feed industry: A review. Reviews in Environmental Science and Bio/Technology. DOI: 10.1007/s11157-020-09557-7

[53] Li H, Shao F, Qiu Y, Ma Y. Solubility, uptake, and translocation of BDE 47 as affected by DOM extracted from agricultural wastes. Environmental Science and Pollution Research. 2019;**26**:19871-19878. DOI: 10.1007/s11356-019-05393-7

[54] Cheng Z, Wang Y, Wang S, Luo C, Li J, Chaemfa C, Jiang H, Zhang G. The influence of land use on the concentration and vertical distribution of PBDEs in soils of an e-waste recycling region of South China. Environmental Pollution. 2014;**191**:126-131. DOI: 10.1016/j.envpol.2014.04.025

[55] Farzana S, Cheung SG, Tam NFY. Effects of aquaculture effluents on fate of 2,2′,4,4′,5-pentabromodiphenyl ether (BDE-99) in contaminated mangrove sediment planted with Kandelia obovata. The Science of the total environment. 2019;**691**:71-79. DOI: 10.1016/j.scitotenv.2019.07.057

[56] Wang Y, Luo C, Li J, Yin H, Zhang G. Influence of plants on the distribution and composition of PBDEs in soils of an e-waste dismantling area: Evidence of the effect of the rhizosphere and selective bioaccumulation. Environmental Pollution. 2014;**186**:104-109. DOI: 10.1016/j.envpol.2013.11.018

[57] Feng J, Shen X, Chen J, Shi J, Xu J, Tang C, Brookes PC, He Y. Improved rhizoremediation for decabromodiphenyl ether (BDE-209) in E-waste contaminated soils. Soil Ecology Letters. 2019;**1**:157-173. DOI: 10.1007/s42832-019-0007-9

[58] Wang S, Wang Y, Song M, Luo C, Li J, Zhang G. Distributions and compositions of old and emerging flame retardants in the rhizosphere and non-rhizosphere soil in an e-waste contaminated area of South China. Environmental Pollution. 2016;**208**:619-625. DOI: 10.1016/j.envpol.2015.10.038

[59] Wang S, Wang Y, Luo C, Li J, Yin H, Zhang G. Plant selective uptake of halogenated flame retardants at an e-waste recycling site in southern China. Environmental Pollution. 2016;**214**:705-712. DOI: 10.1016/j.envpol.2016.04.071

[60] Tian M, Chen S-J, Wang J, Luo Y, Luo X-J, Mai B-X. Plant uptake of atmospheric brominated flame retardants at an E-waste site in southern China. Environmental Science &

Technology. 2012;**46**:2708-2714. DOI: 10.1021/es203669n

[61] Huang H, Zhang S, Christie P, Wang S, Xie M. Behavior of decabromodiphenyl ether (BDE-209) in the soil-plant system: Uptake, translocation, and metabolism in plants and dissipation in soil. Environmental Science & Technology. 2010;**44**:663-667. DOI: 10.1021/es901860r

[62] Jian K, Zhao L, Ya M, Zhang Y, Su H, Meng W, Li J, Su G. Dietary intake of legacy and emerging halogenated flame retardants using food market basket estimations in Nanjing, eastern China. Environmental pollution (Barking, Essex: 1987). 2020;258:113737. DOI: 10.1016/j.envpol.2019.113737

[63] Xiang L, Song Y, Bian Y, Liu G, Herzberger A, Gu C, Jiang X, Wang F. Manure amendment reduced plant uptake and enhanced rhizodegradation of 2,2′,4, 4′-tetrabrominated diphenyl ether in soil. Biology and Fertility of Soils. 2018;**54**:807-817. DOI: 10.1007/s00374-018-1304-7

[64] Wu J, Yi Y, Fang Z, Tsang EP. Effects of biochar on phytotoxicity and translocation of polybrominated diphenyl ethers in Ni/Fe bimetallic nanoparticle-treated soil. Environmental science and pollution research international. 2018;**25**:2570-2579. DOI: 10.1007/s11356-017-0627-5

[65] Kuch B, Rupp S, Fischer K, Kranert M, Metzger JW. Untersuchungen von Komposten und Gärsubstraten auf organische Schadstoffe in Baden-Württemberg: Abschlussbericht. Karlsruhe: Landesanstalt für Umwelt Baden-Württemberg (LUBW); 2007

[66] Teuten EL, Rowland SJ, Galloway TS, Thompson RC. Potential for plastics to transport hydrophobic contaminants. Environmental Science & Technology. 2007;**41**:7759-7764. DOI: 10.1021/es071737s

[67] Xu P, Ge W, Chai C, Zhang Y, Jiang T, Xia B. Sorption of polybrominated diphenyl ethers by microplastics. Marine Pollution Bulletin. 2019;**145**:260-269. DOI: 10.1016/j.marpolbul.2019.05.050

[68] Bao L-J, You J, Zeng EY. Sorption of PBDE in low-density polyethylene film: implications for bioavailability of BDE-209. Environmental Toxicology and Chemistry. 2011;**30**:1731-1738. DOI: 10.1002/etc.564

[69] Ma J, Zhang Q, Chen F, Zhu Q, Wang Y, Liu G. Remediation of PBDEs-metal co-contaminated soil by the combination of metal stabilization, persulfate oxidation and bioremediation. Chemosphere. 2020;**252**:126538. DOI: 10.1016/j.chemosphere.2020.126538

[70] Yao Y, Wang B, He Y, Wang L, Corvini PF-X, Ji R. Fate of 4-bromodiphenyl ether (BDE3) in soil and the effects of co-existed copper. Environmental pollution (Barking, Essex: 1987). 2020;261:114214. DOI: 10.1016/j.envpol.2020.114214

[71] Yan Y, Ma M, Zhang W, Ma W, Li M, Yan L. Effect of elevated nitrate on biodegradation of pentabromodiphenyl ether (BDE-99) and change in microbial communities during groundwater recharge with tertiary-treated municipal wastewater. International Biodeterioration & Biodegradation. 2017;**124**:128-137. DOI: 10.1016/j.ibiod.2017.03.022

[72] Pi N, Wu Y, Zhu HW, Wong YS, Tam NFY. The uptake of mixed PAHs and PBDEs in wastewater by mangrove plants under different tidal flushing regimes. Environmental Pollution. 2017;**231**:104-114. DOI: 10.1016/j.envpol.2017.07.085

[73] Lu M, Zhang Z-Z. Phytoremediation of soil co-contaminated with heavy metals and deca-BDE by co-planting of

Sedum alfredii with tall fescue associated with *Bacillus cereus* JP12. Plant and Soil. 2014;**382**:89-102. DOI: 10.1007/s11104-014-2147-0

[74] Chen F, Zeng S, Ma J, Li X, Zhang S, Zhu Q. Interactions between decabromodiphenyl ether and lead in soil-plant system. Chemosphere. 2019;**236**:124406. DOI: 10.1016/j.chemosphere.2019.124406

[75] Li H, Li X, Xiang L, Zhao HM, Li YW, Cai QY, Zhu L, Mo CH, Wong MH. Phytoremediation of soil co-contaminated with Cd and BDE-209 using hyperaccumulator enhanced by AM fungi and surfactant. The Science of the total environment. 2018;**613-614**:447-455. DOI: 10.1016/j.scitotenv.2017.09.066

[76] Li H, Huang WX, Gao MY, Li X, Xiang L, Mo CH, Li YW, Cai QY, Wong MH, Wu FY. AM fungi increase uptake of Cd and BDE-209 and activities of dismutase and catalase in amaranth (Amaranthus hypochondriacus L.) in two contaminants spiked soil. Ecotoxicology and environmental safety. 2020;**195**:110485. DOI: 10.1016/j.ecoenv.2020.110485

[77] Li Y, Chiou CT, Li H, Schnoor JL. Improved prediction of the bioconcentration factors of organic contaminants from soils into plant/crop roots by related physicochemical parameters. Environment international. 2019;**126**:46-53. DOI: 10.1016/j.envint.2019.02.020

[78] Collins CD, Finnegan E. Modeling the plant uptake of organic chemicals, including the soil-air-plant pathway. Environmental Science & Technology. 2010;**44**:998-1003. DOI: 10.1021/es901941z

[79] Hwang J-I, Lee S-E, Kim J-E. Comparison of theoretical and experimental values for plant uptake of pesticide from soil. PloS one. 2017;**12**:e0172254. DOI: 10.1371/journal.pone.0172254

[80] Briggs GG, Bromilow RH, Evans AA. Relationships between lipophilicity and root uptake and translocation of non-ionised chemicals by barley. Pesticide Science. 1982;**13**:495-504. DOI: 10.1002/ps.2780130506

[81] Bintein S, Devillers J, Karcher W. Nonlinear dependence of fish bioconcentration on n-octanol/water partition coefficient. SAR and QSAR in environmental research. 1993;**1**:29-39. DOI: 10.1080/10629369308028814

[82] Meylan WM, Howard PH, Boethling RS, Aronson D, Printup H, Gouchie S. Improved method for estimating bioconcentration/bioaccumulation factor from octanol/water partition coefficient. Environmental Toxicology and Chemistry. 1999;**18**:664-672. DOI: 10.1002/etc.5620180412

[83] Nendza M. QSARs of bioconcentration - validity assessment of log P/log BCF correlations. Weinheim: VCH; 1991. DOI: 10.13140/2.1.1837.5364

[84] Liagkouridis I, Cousins AP, Cousins IT. Physical-chemical properties and evaluative fate modelling of 'emerging' and 'novel' brominated and organophosphorus flame retardants in the indoor and outdoor environment. The Science of the total environment. 2015;**524-525**:416-426. DOI: 10.1016/j.scitotenv.2015.02.106

Chapter 5

Flame Retardant Treatments of Nylon Textiles: A Shift towards Eco-Friendly Approaches

Chanchal Kumar Kundu, Zhiwei Li, Lei Song and Yuan Hu

Abstract

Among the synthetic polymeric textiles, Nylon (Polyamide) textiles (Nylon66/ Nylon 6) are one of the most widely used materials, especially as apparel and industrial uses for their excellent properties, namely higher strength and good wear resistance. Unfortunately, due to their organic structures, they show the relative ease of burning, which poses a great risk to fire. For the flame retardant (FR) treatment of nylon textiles, several strategies have been developed throughout the years and the earlier studies show the enormous uses of petroleum-based flame retardant compounds via energy intensive application methods. However, the rapid improvement in living standards as well as the recent call for a reduction of environmental impacts during manufacturing and use have been pushed researchers to come up with environmentally benign chemistries and processes. Therefore, the challenges in search of the most sustainable, efficient and durable flame retardant treatments for nylon textiles still remain as a hot topic to be addressed. This chapter discusses the eco-friendly approaches that have been taken in escalating the fire performance of these novel nylon textiles, especially focusing on the applied compounds and the application techniques along with the durability issues of such applications.

Keywords: nylon textiles, flame retardancy, bio-based flame retardant, hybrid application, eco-friendliness

1. Introduction

The consequences of fire related incidents leave a serious impact on human life and its property as well. Frequently, these fires are ignited from the polymeric materials, including textiles we use in our daily life as in consumer goods, home furnishings, transportation, apparel and protective clothing, etc. The world fire statistics reveal that the textiles and upholstered furniture are the first item to be ignited by small flames such as cigarettes and candles, thus producing the most common fire [1, 2]. Meanwhile, the textile materials release smoke and toxic gases while going through the combustion process, which in turn limit the evacuation in fire disasters and worsen the fire scenario [3, 4]. In addition, these fires cause huge economical losses as well as deaths of human beings [5, 6]. Thus, the researcher is making continual efforts to investigate the flammability of textiles in order to improve their fire performance. On the other hand, the growing concern over the sustainability issues infer that fire retardants should leave a low impact on health and the environment during the entire life cycle including recycling and disposal. It also demands to come up with significant researches with a focus on alternate flame retardant chemistries and methodologies, including the use of more environmentally benign raw materials and eco-friendly approaches in the synthesis and application of new flame retardants.

Nylon is the oldest man-made fiber (MMF) among the synthetic textiles, which remains as an important fiber in the synthetic fiber community till the date. Initially Nylon, also in the name of polyamide, is developed for a limited number of end uses; however, these days, the fibers belonging to the nylon/polyamide group share a big market, from regular apparels to technical textiles. For example, carpet is a significant application for nylon and accounts for 17.5 percent of total usage globally. Other applications of nylon include airbag, heavy-duty tires, intimate apparel, military apparel, sheer hosiery and swimwear, etc. [7]. Among, different types of polyamides, polyamide 66 (Nylon 66) and polyamide 6 (Nylon 6) represents one of the most used technical fibers. Both of them possess almost similar physical properties, namely high mechanical properties (tensile strength is higher than that of wool, silk, rayon or cotton), high chemical stability, high melting point, resistance to shrinkage and abrasion [8]. However, like other common textiles these fibers are also flammable due to their organic structure; alongside they also show serious dripping. Thus, the nylon textiles cannot meet industrial and civil requirements in many cases, which ultimately limit their uses in the mentioned sectors [9–11].

2. General combustion behaviors of textiles and strategies of flame retardant mechanism

In general, combustion of a typical polymer substrate happens in contact of a fire source and in the presence of air or oxygen. Prior to the combustion process, the textile materials degrade thermally, while some of the degraded species turn into combustible volatiles and serially, in the presence of oxygen, they kindle the flame. In a logical way, while the heat generation exceeds the threshold to sustain the combustion process, the excessive heat transmitted to the textile material, usually accelerates the degradation process and form a self-sustaining combustion cycle as presented in **Figure 1** [7]. In line, we also need to study the mechanism of action of various flame retardants on textiles to evaluate a particular flame-retardant system

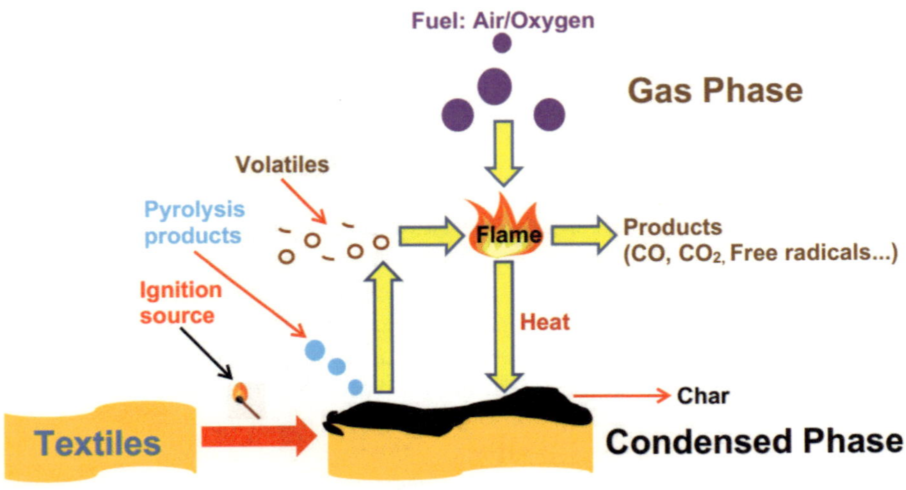

Figure 1.
Combustion cycle of a typical textile material. [7], Copyright 2020. Reproduced with permission from Elsevier ltd.

to a specific polymer. In general, the common flame retardant mechanism can be divided into two different modes of action likely (1) gas or vapor phase mechanism and (2) solid or condensed phase mechanism that break the polymer combustion cycle (see **Figure 1**). The gas phase flame retardant mechanism normally does not change the thermal decomposition of a polymeric substrate, whereas the flame-retardant compound goes for decomposition in the presence of heat and generates free radicals. These free radicals either combine with atmospheric oxygen/air through complex reactions or capture the radicals released from the polymer substrate to quench the total combustion process [12]. Meanwhile, in the condensed phase mechanism, combustion process brings structural changes in the polymer substrate via promoting polymer cross-linking to form a carbonaceous char onto the material surfaces. In line, these char residues make insulation between the polymer substrate and the flaming zone to curb the creation of new fuel and stop further burning. Another form of the condensed phase mechanism involves an additive, creating a physical barrier via releasing water during heating, which ultimately cools down the surface and slows the heat and mass transfer [13].

Briefly, a typical flame retardant (FR) compound inhibits the flammability of a polymeric/textile material in several ways mentioned as follows [14]: (i) FR compound can minimize the generation of heat to retard the combustion process, (ii) it can alter the pyrolysis pathway via lowering the generation of flammable volatiles while favoring the char formation to limit the heat and mass transfer via creating an insulation layer between the textile material and the fire source, (iii) in some extent, it can release water vapor as a byproduct to dilute the concentration of available oxygen and flammable volatiles to lower the heat flow back to the textile material, and (iv) it can release flame inhibitors in the gas phase (i.e., chlorinated, brominated and phosphorus species) to quench the intensity of combustion via capturing flammable radicals.

3. Combustion and thermal behaviors of nylon textiles

To come up with a suitable flame retardant approach along with an applicable flame-retarding agent for nylon textiles, it is thus needed to understand their thermal and flammability behaviors. In general, the polymeric materials release gases like CO, NO_2 and HCN, etc. when they are burned [15] and it is also observed that the evolution of CO differs from fiber to fiber. Meanwhile, polyamide fibers show self-extinguishing behavior due to its extensive shrinkage and dripping in combustion [16]. During burning, polyamide ignites with molten droplets and drip away from the flame; most of the heat is carried away with the droplet, making the material self-extinguishing. However, if the molten droplets burn continuously, this will encourage a greater fire hazard and pose a secondary fire risk. In inert atmosphere and at a higher temperature range (i.e., above 300°C), the main decomposition products released by polyamides are about 95% non-volatiles and the remaining volatile compounds mainly consist of CO_2, CO, water, ethanol, benzene, cyclopentanone, ammonia, others aliphatic and aromatic hydrocarbons, etc. [17, 18]. However, in air atmosphere and at temperatures below 200°C, the degradation pattern of polyamide is different [17] where the volatile products to be likely water 52%, CO_2 33%, CO 12%, and methanol, formaldehyde and acetaldehyde are around 1% each. Moreover, the pyrolysis process also causes de-polymerization of its structure [19]. The suggested oxidative decomposition mechanism of polyamide structure is given in **Figure 2** [20].

From **Figure 2**, it is assumed that like the oxidative degradation of hydrocarbons, the oxygen molecule initiates the chain process of oxidation of polyamides (**Eq. 1**). At first, hydrogen atom will be abstracted and subsequently, either

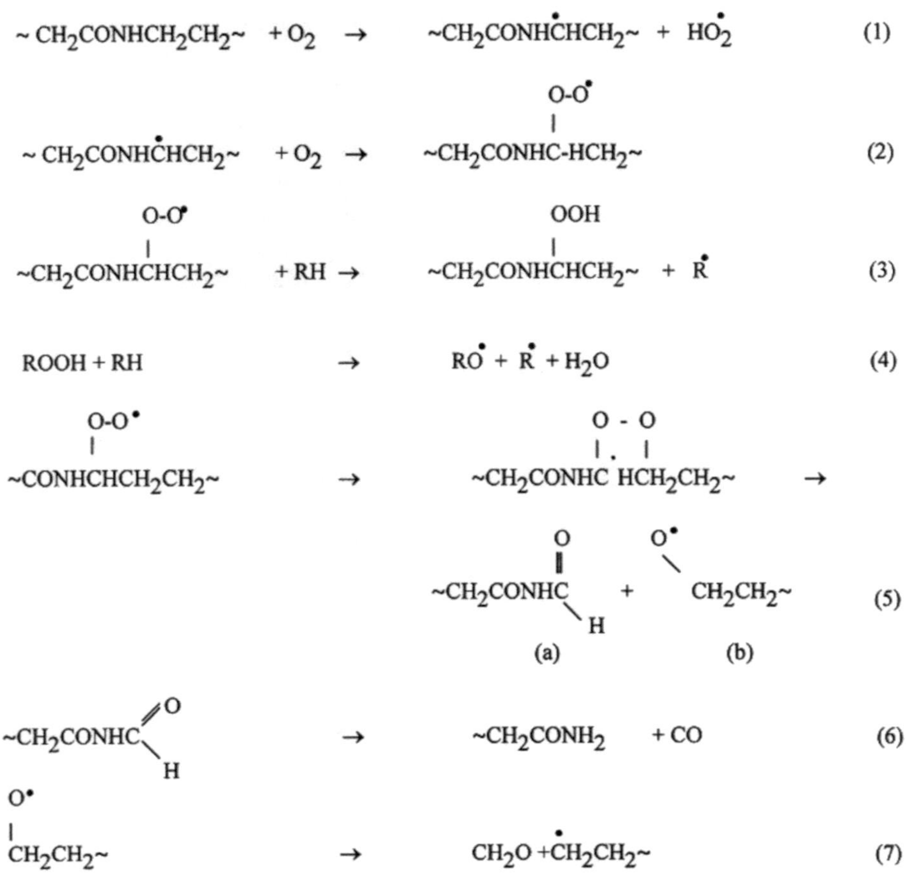

Figure 2.
Possible oxidative decomposition mechanism for polyamides [20].

peroxide radical or a hydroperoxide will be formed (**Eq. 2 and 3**). Later, with the decomposition of hydroperoxide, water will be formed (**Eq. 4**). The formation of water can lead to the hydrolysis of the polymer and thus, on decarboxylation, CO_2 will be produced. Apart from the decomposition of peroxide, the peroxide radicals may also break down in the degradation process. These radicals may also go through isomerization via making reaction with the free valence of the adjacent C-C bond. Ultimately, this kind of isomerization as well as the breakdown of peroxide radicals causes the collapse of molecular chains to form a molecule with a terminal aldehyde group (a) and a radical (b) (**Eq. 5**). Afterwards, these aldehyde groups go for further decomposition to the form CO (**Eq. 6**), while the radicals (b) cause the rupture of the C-C bond to form a secondary C-O bond and thus, in turn direct the creation of formaldehyde (**Eq. 7**).

4. Flame retardant treatment of nylon textiles

Formerly, numerous approaches are considered to make the nylon textiles flame retardant, especially at a fiber stage. Of them, strategies like application of fire retardant additives and co-monomers during polymerization and in the spinning dope, development of inherently fire retardant fibers and fiber blending are notable [21]. However, due to the possibility of polymer degradation and leaching of flame retardant compounds at a fiber stage finishing, a kind of topical finishing or post-treatments are commonly proposed for conferring the flame retardancy of nylon textiles in the recent time [22]. Meanwhile, the earlier applications, mostly receive the attention of halogenated species [23, 24] and petro-based flame retardant compounds containing phosphorus, nitrogen [25, 26] and sulfur elements where the uses of bio-derived flame retardants are rarely seen.

Though the nylon textiles (nylon 66/nylon 6) possess almost similar chemical, physical, thermal and fire behaviors, they belong to a different chemical structure. Thus, they have different surface functional groups and a varied level of crystallinity, which ultimately affects the processing of these textiles during finishing. As we know that nylon 66 is made from two monomers namely, hexamethylenediamine and adipic acid; each containing 6 carbon atoms and carries amino and carboxylic acid as functional groups. While, nylon 6 is synthesized from a single monomer namely caprolactam (also called ε-caprolactam) having 6 carbons via ring-opening polymerization and only carries amino functional groups.

In the following section, the very common flame retardant finishes applied to nylon 66/nylon 6 textiles both at the fiber and fabric stage are discussed.

4.1 Flame retardant (FR) treatment of nylon 66 (polyamide 66) textiles

4.1.1 FR treatment at fiber stage

The fiber stage flame retardant finishing of Nylon 66 textiles is usually done using polymerization, electrospinning, mixing or blending techniques. In some applications, halogen free flame retardant compounds, namely a mixture; comprising of phosphorus and boron esters via ionic introduction [27] and melamine cyanurate (MCA) via in situ polymerization [28] are considered. While, a kind of condensation polymerization using nylon salt and 9,10-dihydro-9,10-oxa-10-phosphaphenanthrene-10-oxide (DOPO)-based flame retardant working in gas phase is also exploited [29] to prepare the flame retardant nylon 66 fibers. Recently, a hybridized flame retardant compound [30], namely a kind of encapsulated nanoscale graphene with red phosphorus is used to prepare core-shell structured polyamide 66 nanofiber via a coaxial electrospinning method. However, till the date, the application of fully green compounds as a flame retardant is still absent in preparing the flame retardant nylon 66 fibers. The main results of these applications are presented in **Table 1.**

4.1.2 FR treatment at fabric stage

The flame retardant treatment of nylon 66 textiles at the fabric stage is usually done in the form of surface coatings via numerous techniques, namely pad-dry-cure, UV grafting, plasma deposition, layer-by-layer assembly and sol–gel process [31–34]. In earlier applications, petro-based compounds are widely used, however, with the growing concern of adverse impacts posed by petro-based flame retardants, researchers are paying attention on the green flame retardants and in some extent, semi-green application approaches are also seen in the recent time.

4.1.2.1 Application of fully bio-based flame-retardants

The flame retardant treatment of nylon textiles using fully bio-based compounds are yet to be in extensive uses as a very few work focuses on it. For example, Kundu et al. [35] consider all green compounds, namely bio-derived chitosan (CS), phytic acid (PA) and oxidized sodium alginate (OSA) for the first time ever to improve the flame retardancy of nylon 66 textiles. These coatings with a 10 to 15 quadra-layer (QL; deposition of consecutive four layers constructs a single quadra-layer) (see **Figure 3**) depositions stop the melt dripping of fabric materials.

Substrate	FR compounds	Application technique	Main results	Eco-friendliness
Nylon 66 fiber [27]	Phosphorus and boron esters	Open air grafting (cold)/ultraviolet irradiation grafting (hot) and hot vacuum oven grafting	Improved charring by about 25.67%	No
Nylon 66 fiber [28]	Melamine cyanurate (MCA)	In-situ polymerization	LOI upto 31.5% and UL94 V-0 rating in VBT	No
Nylon 66 fiber [29]	DOPO-based flame retardant	Condensation polymerization	LOI upto 32.9% and UL94 V-0 rating in VBT	No
Nylon 66 fiber [30]	Graphene and red-Phosphorus	Encapsulation	THR and pHRR reduction by about 50.5 and 26.9% respectively	No

Table 1.
Collected results of flame retardant treatment for nylon 66 fiber.

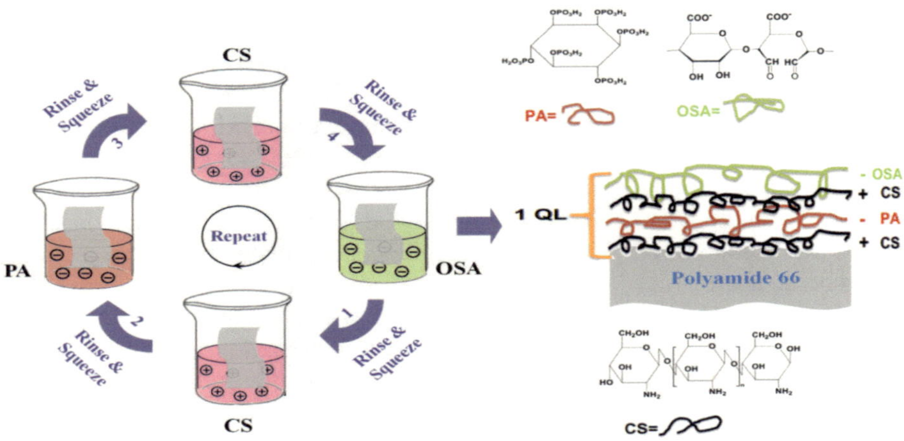

Figure 3.
Schematic representation of Quadra-layer (QL) deposition.

Here, the use of these green polyelectrolytes plays a multiple role as an acid source, a carbon source and a foaming agent and thus, their interactions facilitate the intumescent process in a condensed phase mechanism to control the flame spread process. More importantly, it is found that the use of OSA improves the durability of this multilayered nanocoating via offering covalent interactions. This is the first attempt to use fully renewable and environmentally friendly intumescent LbL coating in improving the flame retardant properties of a thermoplastic textile like nylon 66. In addition, this is also a very first attempt to improve the durability of LbL assembled coatings for nylon textiles against home laundering as the coatings developed via non-covalent forces (i.e., ionic interaction) show poor wash durability. In line, another environmentally friendly and fully bio-based synthesized compound, namely tannic acid-terephthalate (TAT) (see **Figure 4x**) obtained from tannic acid, is applied onto the nylon 66 fabric surfaces via pad-dry-cure process [36]. This kind of green coating surely imparts excellent flame retardancy to the textile substrate in terms of obtaining self-extinguishing behavior and reduced char length in the

DOI: http://dx.doi.org/10.5772/intechopen.94880

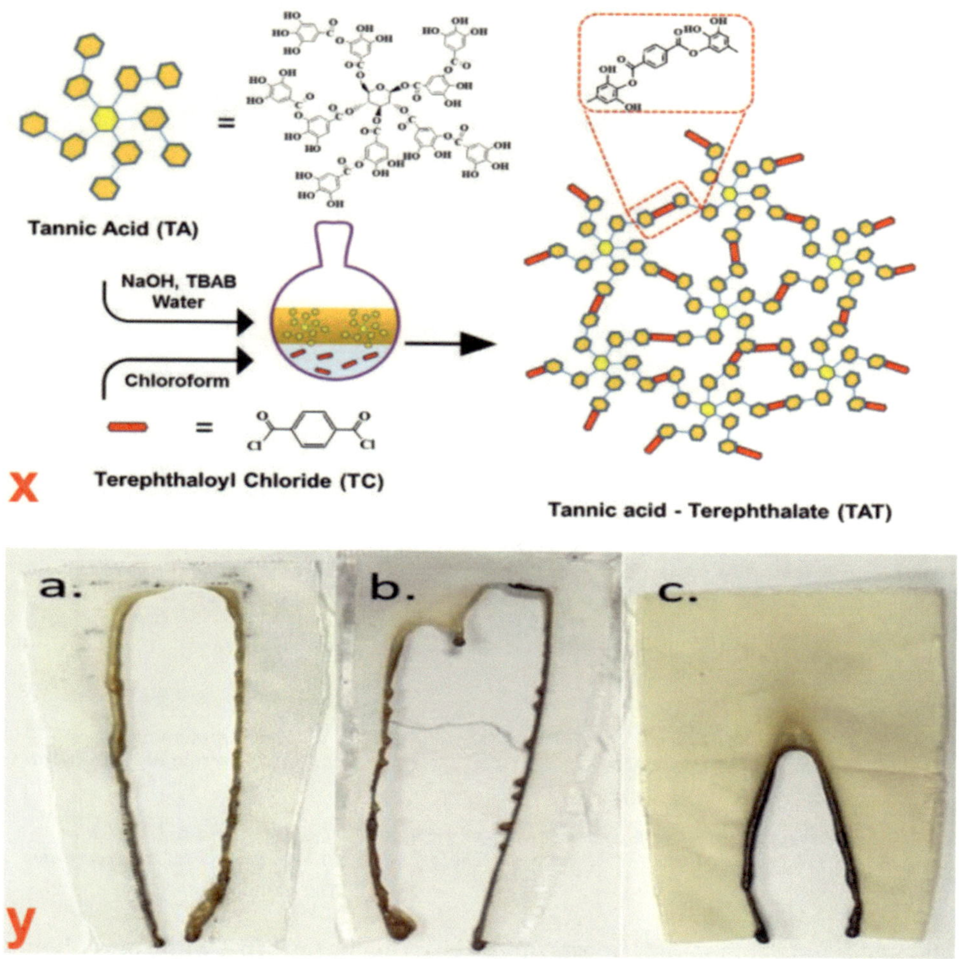

Figure 4.
Proposed reaction scheme of tannic acid and terephthaloyl chloride (top-x) and vertical flame test samples after burning, a. neat nylon 66 fabric, b. TA-coated nylon 66 fabric (after washing), c. TAT-coated nylon 66 fabric (after washing) (bottom-y) [36], copyright 2018. Reproduced with permission from Elsevier ltd.

vertical flame test (see **Figure 4y**). Here, the cross-linked phenol species belonging to TAT structure helps to build up charring in the solid phase and exhibits a positive impact in enhancing flame retardancy of nylon substrate. Just a very thin layer of such coating (i.e., only add-on of 6 wt.%) helps to obtain self-extinguishing behavior and brings a considerable decrease (i.e., 20–25%) in heat release rate. The main results of these applications are summarized in **Table 2.**

4.1.2.2 Application of hybrid flame-retardants

To address the environmental issues, researchers equally put focus on the numerous hybrid application techniques, namely they consider the combined use of both the bio-derived and petro-based compounds, even in some applications, they opt to blend the mineral with the bio/petro-based compounds in a semi eco-friendly application. For example, Kundu et al. [37] combine an inorganic boron compound like sodium tetraborate with the bio-derived compounds like chitosan (CS) and phytic acid (PA) to simultaneously improve the flame retardant properties of PA66 textiles and to stabilize the ionically assembled coatings (see **Figure 5**). Here, the modification of nylon 66 textile with chitosan, phytic

Substrate	FR compounds	Application technique	Main results	Eco-friendliness
Nylon 66 fabric [35]	Chitosan, phytic acid and oxidized sodium alginate	LbL assembly	A maximum of 24% reduction in pHRR	Full
Nylon 66 fabric [36]	Tannic acid terephthalate (TAT)	Pad-dry-cure	Improved charring and self-extinguishing behavior	Full

Table 2.
Collected results of flame retardant treatment for nylon 66 fabric.

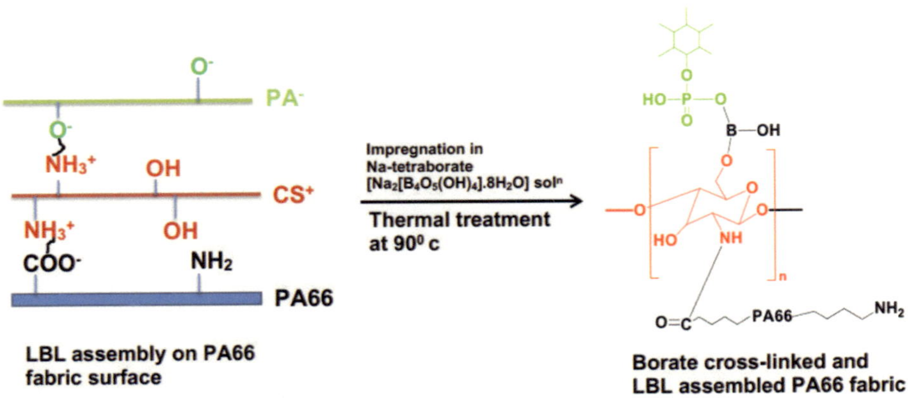

Figure 5.
Schematic illustration for the interactions among chitosan, phytic acid, borate compound and PA66 fabric during the LbL deposition and after the thermal treatment [37], copyright 2018. Reproduced with permission from Elsevier B.V.

acid and borate compounds boosts up the charring of the same while providing enhanced stability of char residues. In addition, the borate induced cross-linking offers added stability of as-prepared coating in detergent solution.

Presently, this hybrid methodology has been expanded further in several works where Kundu et al. [38] come up with the hybridization of some alkoxy silane based compounds, namely (3-Aminopropyl) triethoxysilane (APTES) and boron doped APTES sol solutions with some bio-originated compounds like chitosan and phytic acid in the hybrid application (i.e., LbL deposition/sol–gel process). Here, the simultaneously APTES and B-d-APTES sols treated and LbL deposited fabrics stop the melt-dripping in the vertical burning test. Meanwhile, the boron doped APTES sol treated fabrics reveal improved charring and thermal stability in TG analysis. It is speculated that the phytic acid, being a phosphorus compound, catalyzes the charring while the silicon and boron compounds endow with excellent shielding to these char residues. Thus, an added level of char yield% is experienced and finally, this thermally stable char residues impart enhanced flame retardancy to the nylon textile. In addition, this kind of hybrid application, namely LbL assembly (i.e., 5 bi-layers deposition) and simultaneous sol–gel treatment offers durability to the applied finish against home laundering. In another application, Kundu et al. [39] introduce a hybrid methodology via using a chitosan derivative, namely phosphorylated chitosan (PCS) along with an alkoxy-silane like (3-aminopropyl) triethoxysilane (APTES). Here, the PCS is grafted onto the surface of nylon 66 fabrics via UV-induced grafting polymerization and subsequently, these grafted fabrics are further modified by APTES through sol–gel process in order to form a cross-linked coating (see **Figure 6**). The obtained results indicate that the simultaneous PCS

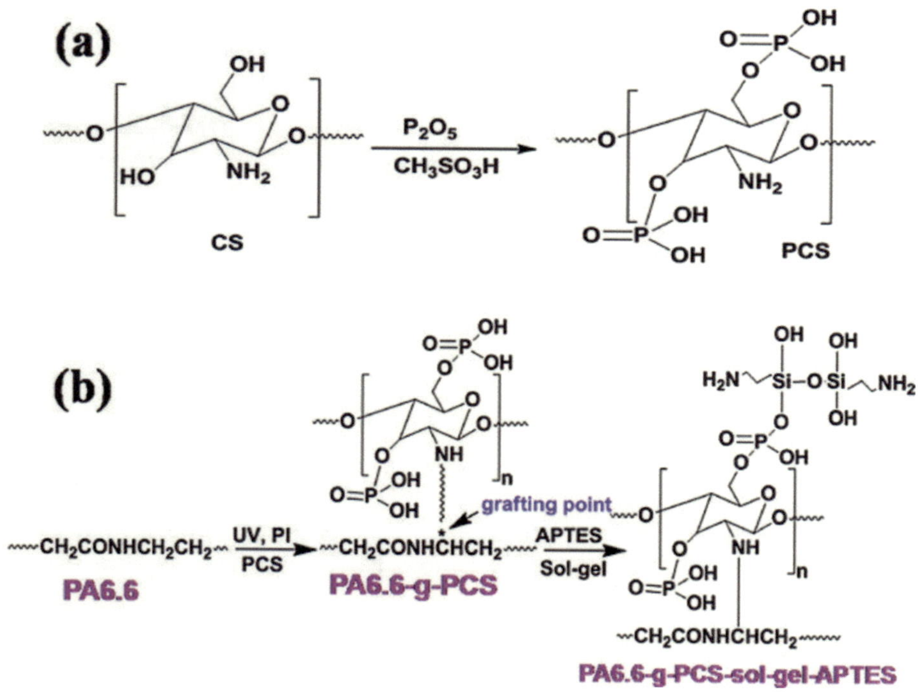

Figure 6.
Synthetic route of PCS (a) and schematic illustration of UV grafting and sol–gel treatment of PA66 fabrics (b) [39], copyright 2017. Reproduced with permission from Elsevier ltd.

grafting and sol–gel treatment stop the melt dripping and improve other flame retardant parameters. The prominent flame retardant properties for the simultaneous PCS and APTES treated PA66 fabrics are due to the joint effect of thermal shielding exerted by the silica and char-forming effect derived from PCS.

Hybridization in the deposition method, namely synchronization of layer by layer (LBL) assembly with the pad-dry-cure technique is seen in preparing a flame retardant hybrid coating for nylon 66 fabric [40]. Here, some bio-derived compounds like chitosan and phytic acid are combined with melamine and graphene oxide (GO). This hybrid application significantly improves the flame retardancy with a considerable improvement in thermal stability and char yield%. Furthermore, the application of a nanocomposite layer prepared from chitosan/graphene oxide on top of the LbL assembled film offers covalent cross-linking to the underneath ionic-assembled coating, which impart considerable durability to the applied finishing to retain flame retardant performance even after 10 laundering cycles. In line, graphene oxide (GO) is functionalized with some plant derived biomolecules namely, lignin (L) and phytic acid (PA) (see **Figure 7**) and used accordingly in modifying the fire performance of nylon 66 fabrics in a one pot deposition [41]. Here, in some formulations, chitosan as a naturally derived charring agent is also considered to enhance the charrability of such finishing further. The as prepared thin film (i.e., add on is less than 10%) barely alters the physical properties (i.e., color, handle and tensile strength, etc.) of treated fabrics. Meanwhile, this kind of application offers better thermal stability and improved fire performance alongside the wash durability of applied finishes. Here, the inclusion of citric acid as a cross-linking agent extends some cross-linking interaction among the applied compounds to retain the flame retardancy even after 5 washing cycles.

Another bio-derived compound, namely soybean protein isolation (SPI) is used with thiourea via a simple pad-dry process to impart flame retardant properties to

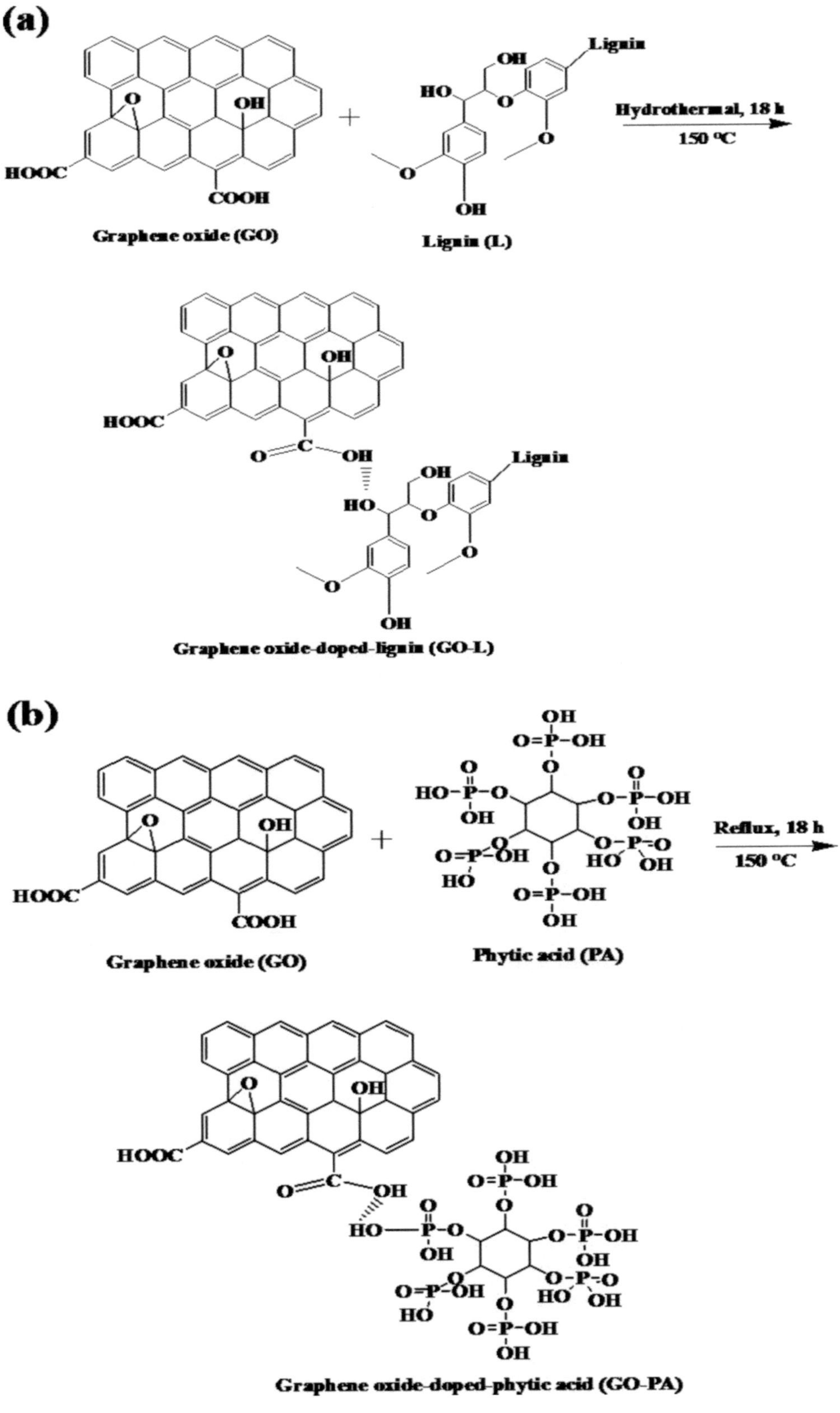

Figure 7.
Synthesis route of graphene oxide-doped-lignin (GO-L) (a) and graphene oxide-doped-phytic acid (GO-PA) (b) [41], copyright 2020. Reproduced with permission from Elsevier B.V.

nylon 66 fabrics [42]. This kind of hybridization in the applied compounds comes up with reduced burning length and no dripping for the modified textiles. Here, it is speculated that these hybrid compounds may release non-combustible gases like H_2S, NH_3 and H_2O in the gas phase and produce compact char residues in the solid phase during burning. In line, these noncombustible gases dilute the fuel/oxygen concentration and the obtained char layers provide insulation between the fire zone and textile substrate to prevent the transfer of heat and oxygen. Therefore, it is perceived that the N, P, S and metal elements in SPI play a vital role in modifying the fire behaviors of polymeric substrates and offer better flame retardancy in a sustainable manner.

In some application, pure chitosan (CS) along with its phosphorus-derivative, namely phosphorylated chitosan (PCS) are used to construct a flame retardant coating for nylon 66 textiles while poly-acrylate sodium (PAS) is accompanied with them as a cross-linking agent [43]. The coating is prepared both in LbL assembly and one pot deposition method to make a comparison between the application method (see **Figure 8**). Here, the LbL assembled coatings possess superior homogeneity in the coating structure over the one pot deposited one. Meanwhile, the LbL assembled coatings with a higher add on% offer an improved LOI value and also bring a considerable

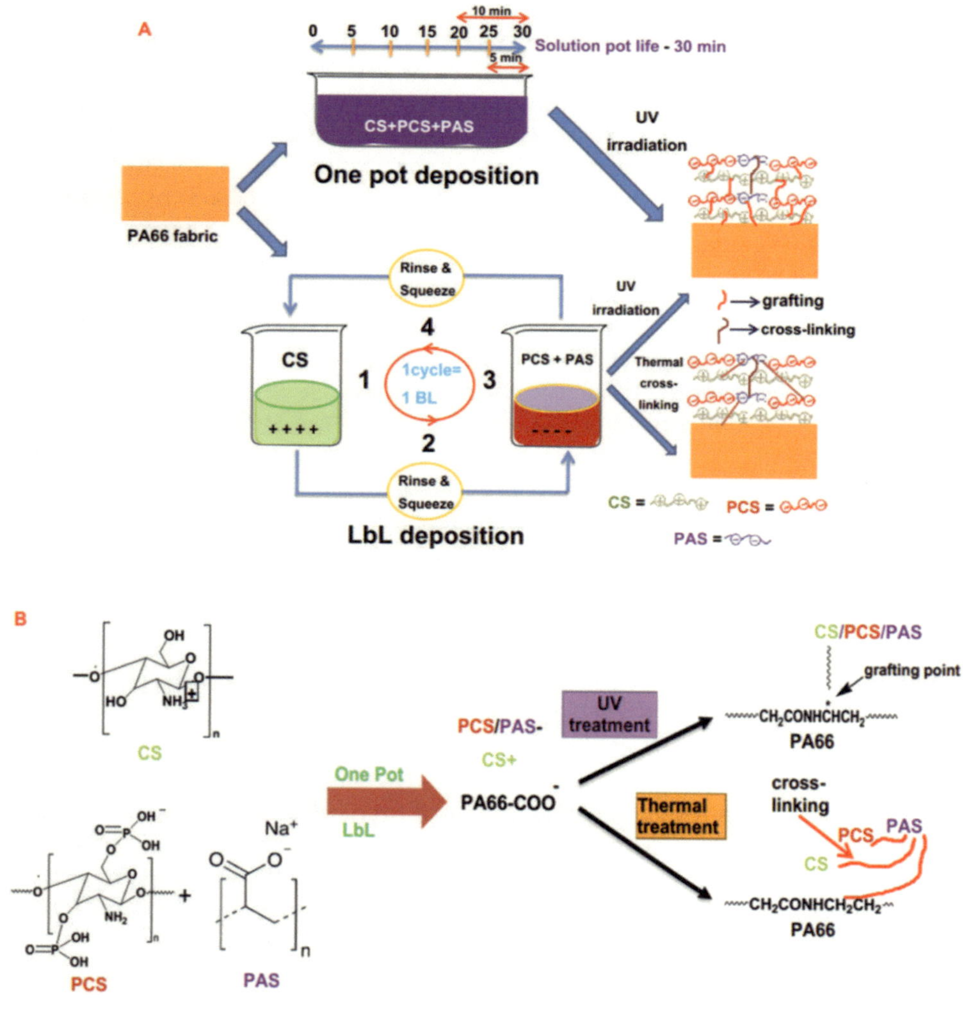

Figure 8.
Schematic representation of the 'one-pot', and LbL deposition of PA 66 fabrics (A) and the mechanism of reactions (B) [43], copyright 2019. Reproduced with permission from Elsevier B.V.

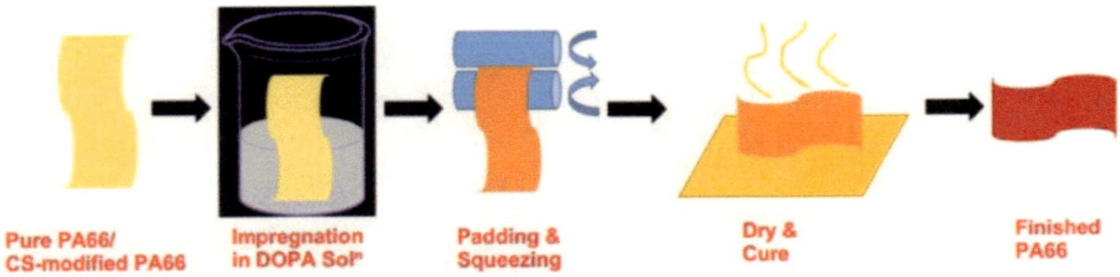

Figure 9.
Schematic representation of DOPA application [44], copyright 2020. Reproduced with permission from Elsevier ltd.

reduction in pHRR for the nylon substrate. In addition, the PAS cross-linking via thermal treatment stabilizes the deposited coatings to a certain extent; as the fabrics prepared in this process retain the UL-94-V1 rating after the 5 times washing in laundering. It is also found that the LbL assembly imparts better flame retardancy than the one-pot deposition technique as a higher weight gain% in LbL deposition offer this superior flame retardant properties irrespective of deposition techniques.

In another application [44], a petro-based P-compound, namely DOPA, a DOPO derivative, is combined with the bio-based chitosan (CS) in a semi bio-based flame retardant application via a simple pad-dry-cure method as shown in **Figure 9**. Here, the nylon 66 modified with chitosan and 10 wt. % of DOPA reveals an excellent flame retardancy in terms of significant reduction in pHRR compared to the nylon 66 modified only with DOPA. It is speculated that the P and N elements available from DOPA and chitosan respectively appear to be involved synergistically to offer enhanced flame retardancy. In line, the co-presence of DOPA and CS ensures a synergistic gas phase and solid phase activity where DOPA exhibits a quenching effect in the gas phase and chitosan takes part in the charring process in the condensed phase.

Even in some applications, several novel nanoparticles (NPs) namely, TiO_2 and SiO_2 are blended with naturally derived phytic acid (PA) and chitosan (CS) via a pad-dry-cure technique [45]. The blended formulation of PA-TiO_2 enhances the limiting oxygen index (LOI) and increases char yield% of treated textiles. Conversely, the introduction of chitosan (CS) into the PA-TiO_2/SiO_2 formulations brings a significant reduction in the peak heat release rate (pHRR). In addition, the formulations containing nanoparticles (TiO_2/SiO_2) along with phytic acid boost up the tensile strength of treated textiles. Here, it is realized that the hybrid complexes of nanoparticles and bio-based compounds can be considered as a big alternative of the traditional halogen and phosphorus containing synthetic compounds. The main results of these applications are presented in **Table 3.**

4.2 Flame retardant (FR) treatment of nylon 6 (polyamide 6) textiles

4.2.1 FR treatment at fiber stage

At the fiber stage finishing, a higher loading of a typical flame retardant is usually considered in imparting a satisfactory level of flame retardancy, which ultimately affects the mechanical properties of treated fibers and causes the spinning process difficult. For example, Coquelle et al. [46] use ammonium sulfamate (AS) as an additive flame retardant for nylon 6. Here, a loading of 5 wt.% AS in PA6 matrix leaves no significant changes since the pure PA6 and modified PA6 fiber come up with an almost similiar fiber diameter (see **Figure 10**). An investigation based on the loading % of AS on the physical properties of modified nylon 6 fibers show that the fibers modified with less than 7 wt. % of AS are quite spinnable without alteration of the mechanical properties. In contrast, a higher loading (i.e., 10 wt. %) makes the material brittle and

Substrate	FR compounds	Application technique	Main results	Eco-friendliness
Nylon 66 fabric [37]	Chitosan, phytic acid and Na-metaborate	LbL assembly/dipping	LOI up to 21.5%, reduction by 31% in pHRR	Partial
Nylon 66 fabric [38]	Chitosan, phytic acid, APTES and B-d-APTES	LbL assembly/sol–gel process	LOI up to 20.6%, 30% reduction in pHRR	Partial
Nylon 66 fabric [39]	P-chitosan (PCS) and APTES	UV-grafting/sol–gel process	LOI up to 24.3%, reduction by 30% in pHRR	Partial
Nylon 66 fabric [40]	Chitosan, phytic acid, melamine, urea and GO	Pad-dry-cure	LOI up to 25%, pHRR reductin by about 56%	Partial
Nylon 66 fabric [41]	Lignin, phytic acid, chitosan and GO	One pot deposition	LOI up to 27%, reduction in pHRR by 25%	Partial
Nylon 66 fabric [42]	Soybean protein isolation (SPI) and thiourea	Pad-dry-cure	LOI up to 25.5%, no dripping	Partial
Nylon 66 fabric [43]	Chitosan, P-chitosan and PAS	One pot deposition/LBL assembly	LOI up to 23%, 25% reduction in pHRR	Partial
Nylon 66 fabric [44]	DOPA and chitosan	UV grafting/ pad-dry-cure	LOI up to 24.5%, reduction in pHRR by 40%	Partial
Nylon 66 fabric [45]	Phytic acid, chitosan, TiO_2 and SiO_2	Pad-dry-cure	LOI up to 24.5%, reduction in pHRR by 25%	Partial

Table 3.
Collected results of flame retardant treatment for nylon 66 fabric.

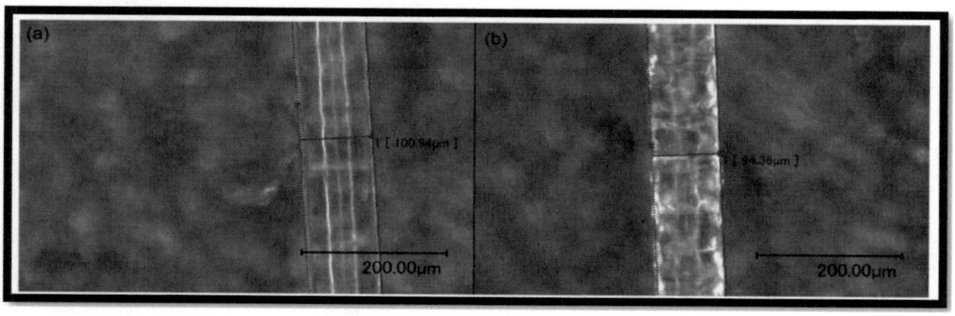

Figure 10.
Optical microscope photographs of PA6 (a) and PA6/AS 5% (b) fibers [46], copyright 2014. Reproduced with permission from Elsevier ltd.

thus, it is impossible to spin it into fibers. This phenomenon exhibits the expected negative impact of higher loading % on the physical properties of synthetic fibers and thus, it is not beneficial to add the flame retardant compounds in the bulk state via additive mixing to impart flame retardant finishing at the fiber stage.

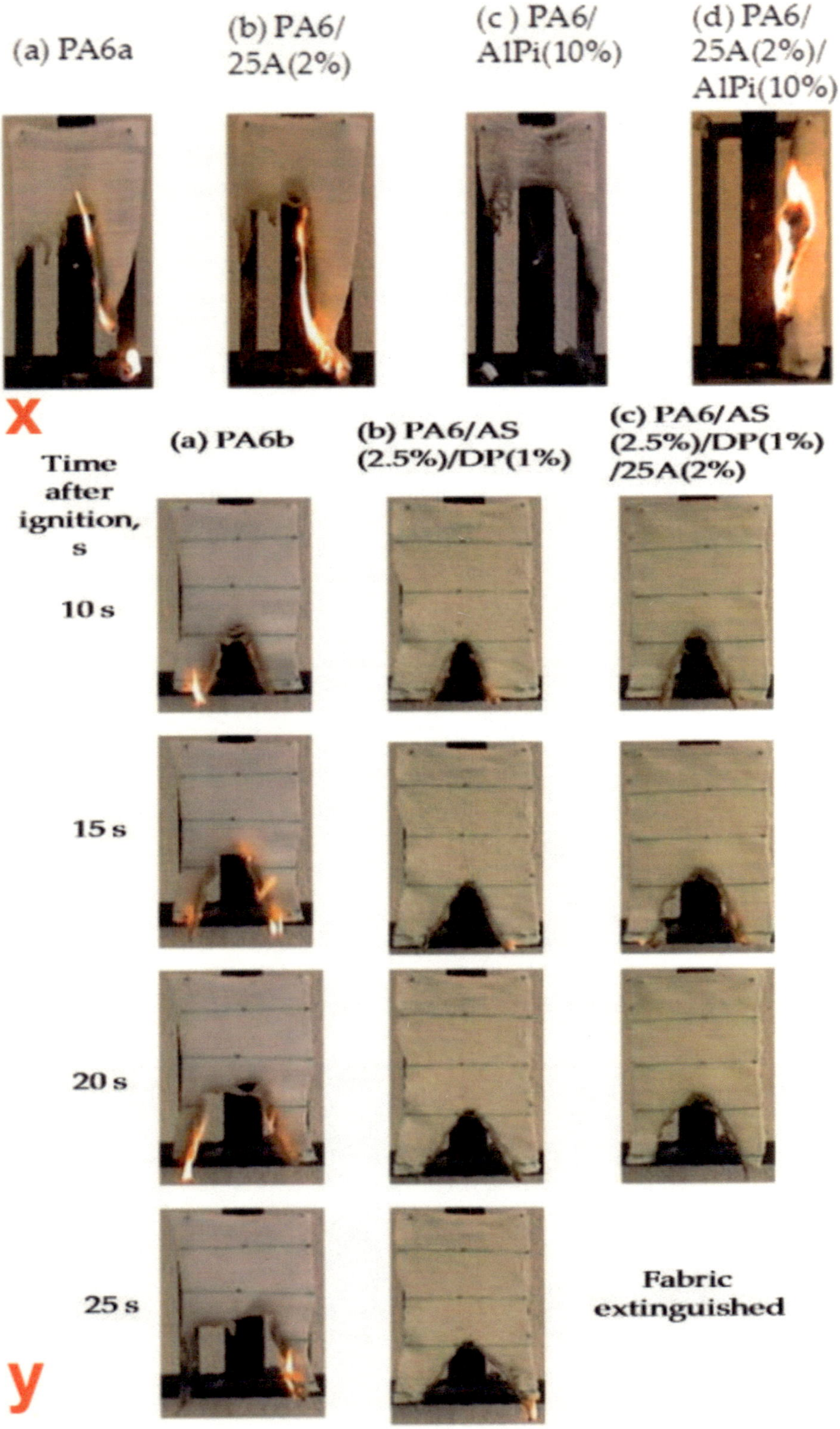

Figure 11.
Selected images of burning PA6 fabrics comprising ALPi and/or Cloisite 25a 40 s after extinction of igniting flame (top-x) and burning and extinguishing behavior of ignited PA6/DP/AS/clay-based fabric samples: 10, 15, 20 and 25 s after extinction of the igniting flame (bottom-y) [47], copyright 2016. Reproduced with permission from www.mdpi.com/journal/polymers.

So, a low loading of additive compounds is preferable (i.e., less than or equal 10 wt. %) to retain spinablity of fibers to make the fabric from them and thus, a kind of trade-off between the loading of FR compounds and mechanical properties of fibers should be maintained in the fire retardant finishing of nylon 6 fibers. This low loading approach has been materialized in some application [47] where two different formulations are considered working in the different flame retardant mechanism (i.e., gas/solid). For example, aluminum diethyl phosphinate (AlPi) working primarily in the gas phase is mixed up with some clay additives like nanoclay, organically modified montmorillonite clay and Cloisite 25A active in the condensed phase. While, in another formulation, ammonium sulphamate (AS)/dipentaerythritol (DP) at 2.5/1 wt. % supposed to be active in the condensed phase, are blended with the same clay-based additives as mentioned earlier. Here, the different formulations come up with different level of flame retardancy as the AlPi-based formulation is incapable to impart self-extinguishability, though the clay-based additives can minimize the burning rate and modify the melt dripping phenomenon. In contrast, the AS/DP-based formulation at a loading of 5.5 wt. % or less offers better flame retardancy and in addition, the clay additives helps to rectify the dripping phenomenon remarkably. From the images of burning fabrics, it is observed that the PA6/AS/DP/clay-containing fabrics (see **Figure 11y**) show definite self-extinguishing characteristics unlike the PA6/AlPi/clay-containing fabric substrates (see **Figure 11x**) and thus, the PA6/AS/DP/clay formulation exhibits potential to be considered in the commercial scale.

Another clay based (i.e., MMT) additive approach is considered [48] to prepare the flame retardant nylon 6 nanocomposite nanofibers. Here, the MMT platelets as a nano-filler is blended with Exolit OP1312; a phosphorus-based non-halogenated additive via electrospinning. In this application, a higher loading of filler material (i.e., nanoclay particles) also affects the electrospinnability of this nanocomposite like other ordinary additive compounds. In another application [49], flame retardant PA6 fiber is developed using a composite formulation, namely via incorporation of melamine cyanurate (MCA) in nylon 6 matrix through in-situ polymerization process. Here, the MCA disperses into the nylon 6 matrix in a homogeneous manner without agglomeration as seen in the SEM images (see **Figure 12**) and thus, the interfacial interaction with the matrix is amplified in many folds. The as prepared composite fiber with a

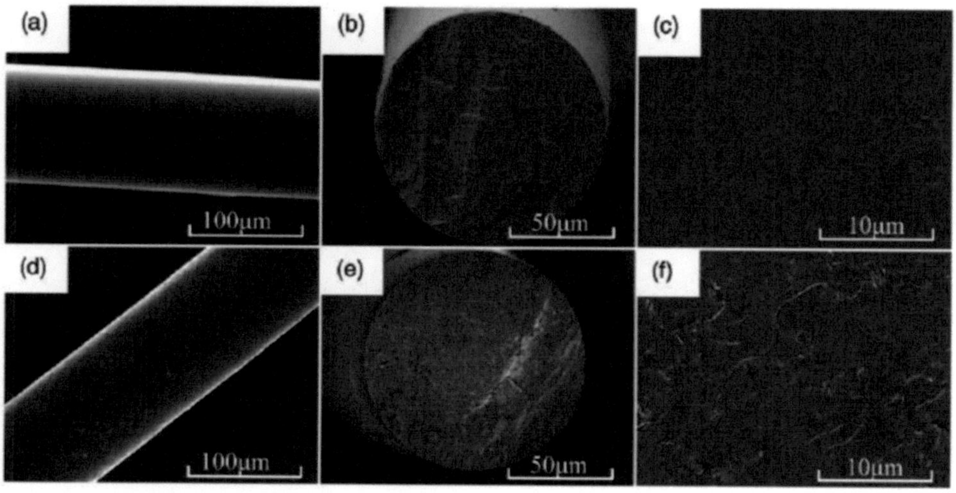

Figure 12.
The scanning electron micrographs of the surface and cross-section of polyamide 6 (PA6) fiber (a), (b), (c) and melamine cyanurate (MCA)/PA-6-8 fiber (d), (e), (f) [49], copyright 2016. Reproduced with permission from sagepub.co.uk/journalsPermissions.Nav.

loading of only 8 wt. % of MCA shows satisfactory flame retardancy. Even, this lower loading % deteriorates the physical properties of the composite fiber as the tenacity at break decreases from 4.85 to 3.11 $cN.dtex^{-1}$, but still meet the requirements of fabric preparation. Moreover, this in-situ polymerization approach opens the way for the preparation of flame retardant nylon 6 fibers with good spinnability.

Apart from these, a phosphorus-based DOPO derivative, namely 9, 10-dihydro-10- [2, 3 di (hydroxycarbonyl) propyl]-10-phosphaphenanthrene-10-oxide (DDP), active in the gas phase, goes through the polycondensation reaction with caprolactam (see **Figure 13**) to develop intrinsically flame retardant nylon 6 fibers [50]. The melt-spun fiber prepared with a loading of 5 wt. % DDP attains a V-0 rating and also obtains a significant increase in LOI value. In addition, this kind of modification does not alter much the spinnability of such fibers in fabric preparation. Here, DDP acts both in the solid and gas phase via escalating charring and producing noncombustible gases respectively to impart fire retardant properties to the modified nylon 6 fibers.

The application of petro-based phosphorus compounds are further extended to develop flame-retarded nylon 6 fibers [51]. Here, at the first stage, co-condensation of ε-caprolactam is carried out via melt-polymerization process and later, a low loading of phosphorus-based FR co-monomer is introduced into the polymer chain with an aim to keep the tensile properties less hampered. Two different phosphorus compounds, namely 3 hydroxyphenylphosphinylpropanoic acid (3-HPP, 1) and 9,10-dihydro-10- [2,3 di(hydroxycarbonylpropyl]-10-phosphaphenanthrene-10-oxide (DDP, 2) are considered (see **Figure 14a**) and finally, they are integrated into the nylon 6 backbones (see **Figure 14b**). Such treatment eases the processing of flame

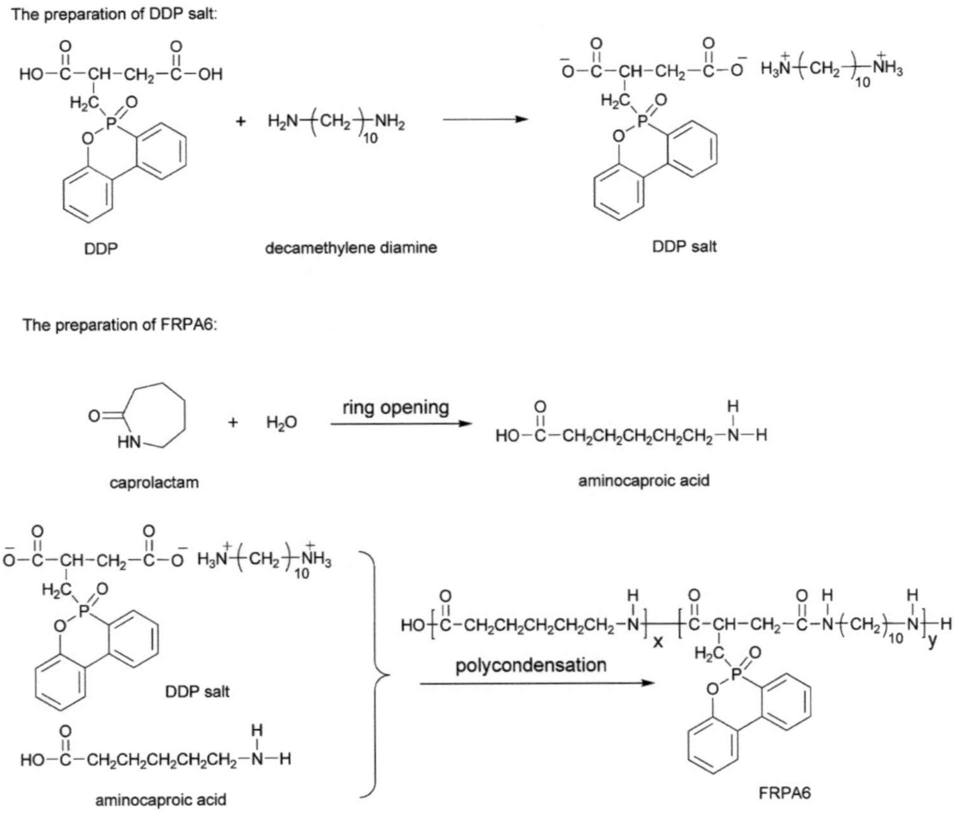

Figure 13.
Preparation process of FR-PA6 [50], copyright 2018. Reproduced with permission from the Royal Society of Chemistry.

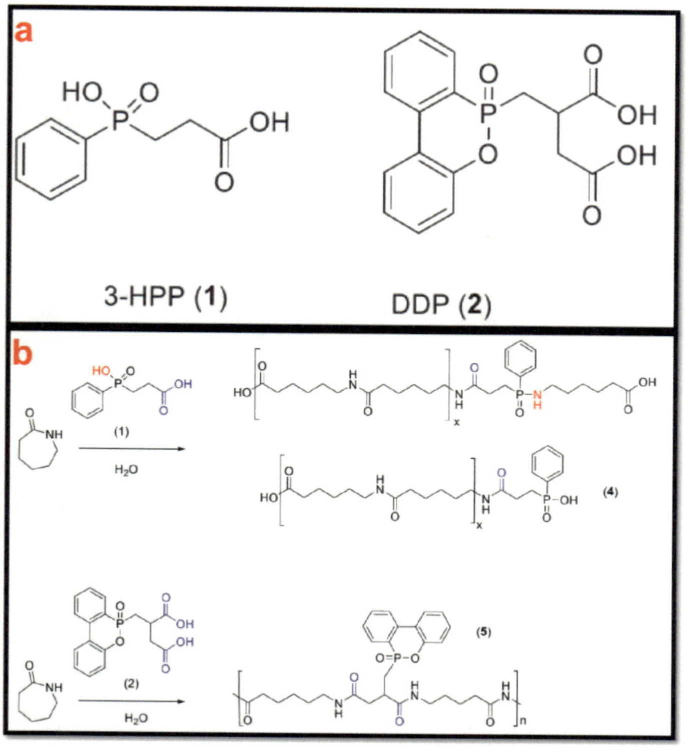

Figure 14.
Chemical formulae of 1 and 2 (top-a) and installation options of the flame retardants into the nylon 6 backbone (bottom-b) [51], copyright 2019. Reproduced with permission from Wiley periodicals, Inc.

retardant nylon 6, especially at the fiber stage. In addition, this kind of facile application rarely alters the tensile properties of nylon substrate, which usually happens in the bulk state application of flame retardant compounds. In line, hexaphenoxycyclotriphosphazene (HPCP) [52], a petro-based phosphorus compound, is considered to modify the flame-retardant properties of nylon 6 fibers (see **Figure 15a**). Here, the nylon 6 chips are blended with the flame retardant compound (HPCP) and fibers are prepared via melt spinning process. It is found that a loading up to 15 wt. % of HPCP does not deteriorate the spinnability of the modified fibers while a slight reduction of tensile properties is observed. Meanwhile, such modification imparts added flame retardancy to the nylon 6 fibers, especially in terms of increased LOI value and improved charring. In addition, the burning behavior is modified as the damaged length and burning time become shortened while dripping rate is minimized and no smoldering effect is observed (see **Figure 15b**). Here, HPCP primarily acts in the condensed phase via producing aromatic char residues along with the release of noncombustible gases like CO, CO_2 and NH_3 in the gas phase.

Apart from the previous techniques, a newly adopted technique, namely electron beam irradiation [53] is also considered in several applications to develop flame retarded nylon 6 fibers. Using this irradiation, some sort of cross-linking is induced to the polymer structure to alter the physical–chemical and thermal behaviors of the same. In some occasion, this irradiation-based cross-linking is carried out in presence of triallylcyanurate (TAC) to avail added benefits. As this kind of cross-linking usually improves the anti-dripping properties of nylon fibers while enhancing the physical properties like breaking strength and elongation at break, especially at a lower dosing. In line, such irradiation is also carried out in the presence of γ-rays along with some sensitizers like trimethallylisocyanurate (TMAIC)

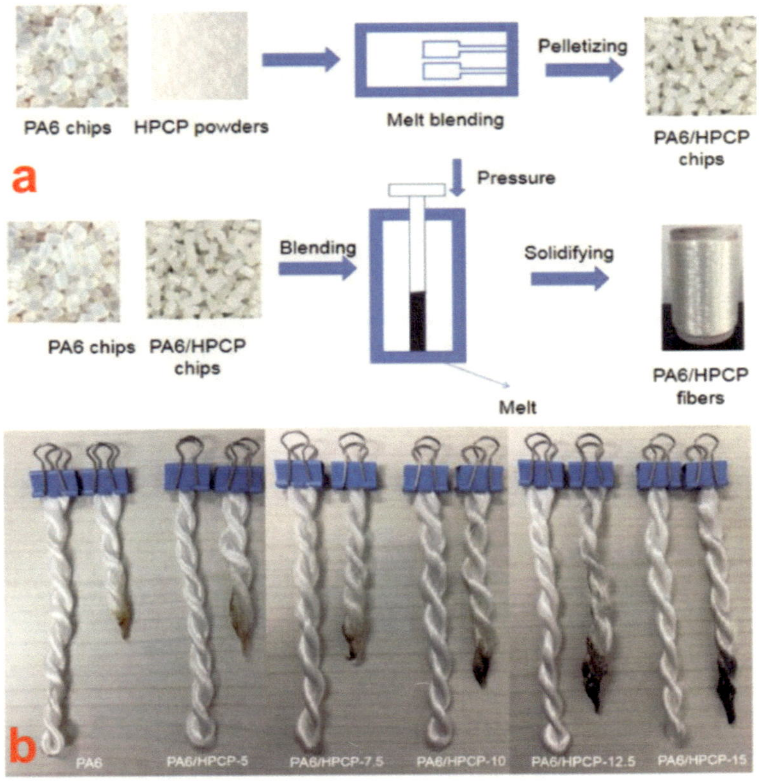

Figure 15.
The preparation process of PA6/HPCP fibers (top-a) and photos of the representative PA6 and PA6/HPCP fiber samples before and after the vertical oriented burning test (bottom-b) [52], copyright 2019. Reproduced with permission from Wiley periodicals, Inc.

and trimethylolpropanetrimethylacrylate (TMPTMA) [54]. In order to overcome the low yield % in cross-linking reaction under the low radiation technique, different types of sensitizers are associated to overcome this shortcoming [55]. Meanwhile, this sensitizer assisted cross-linking alters the nylon 6 fiber structure into a three-dimensional network, which imparts improved anti-dripping properties to the modified fibers. The major results of the above section are summarized in **Table 4.**

4.2.2 FR treatment at fabric stage

A very few attempts have been adopted so far to modify the flame retardant behaviors of nylon 6 textiles at fabric stage. Apaydin et al. [56] first take the initiative to come up with surface finishing of nylon 6 fabrics and introduce layer by layer (LbL) technique in preparing a flame retardant coating using two different types of polyelectrolytes (see **Figure 16x**). One is petro-based organic compound, namely polyallylamine (PAH) (polycation) and the other is an inorganic mineral compound like montmorillonite (MMT) (polyanion). The prepared coatings come up with a homogeneous distribution of coating ingredients with some roughness while in 20 bilayers deposited fabric sample, the MMT platelets lay in a parallel alignment (see **Figure 16y**). And this parallel distribution of MMT platelets ultimately imparts better mechanical and barrier properties to the coating structure. Meanwhile, the as prepared coatings, especially with 10 and 20 bilayers deposition offer enhanced thermal stability of the fabric substrate with a significant reduction in pHRR.

Substrate	FR compounds	Application technique	Main results	Eco-friendliness
Nylon 6 fiber [46]	Ammonium sulfamate (AS)	Additive mixing	30% reduction in pHRR	No
Nylon 6 fiber [47]	Cloisite 25A, aluminum diethyl phosphinate (AlPi), ammonium sulphamate/ dipentaerythritol	Additive mixing	Low burning rate and the lowest level of melt-dripping.	No
Nylon 6 fiber [48]	MMT and Exolit OP1312	Additive mixing	Suppress peak heat release rate and improve charring	Partial
Nylon 6 fiber [49]	ε-caprolactam, adipic acid-melamine salt and cyanuric acid-hexane diamine salt	In-situ polymerization	Effectively suppress the propagation of flame	No
Nylon 6 fiber [50]	Caprolactam and DDP	Melt-polymerization	LOI value up to 33.7%, V-0 rating	No
Nylon 6 fiber [51]	ε-caprolactam, 3-HPP and DDP	Melt-polymerization	LOI up to 35%, good thermal properties	No
Nylon 6 fiber [52]	PA6/hexaphenoxycyclotriphosphazene (HPCP)	Blending technique	LOI up to 28.6%, no smoldering in the vertical burning.	No
Nylon 6 fiber [53]	Triallylcyanurate (TAC)	Electron beam irradiation	Nonvolatile residue increases	No
Nylon 6 fiber [54]	Trimethallylisocyanurate (TMAIC) and trimethylolpropane trimethylacrylate (TMPTMA)	γ-irradiation	Improved anti-dripping performance	No

Table 4.
Collected results of flame retardant treatment for nylon 6 fiber.

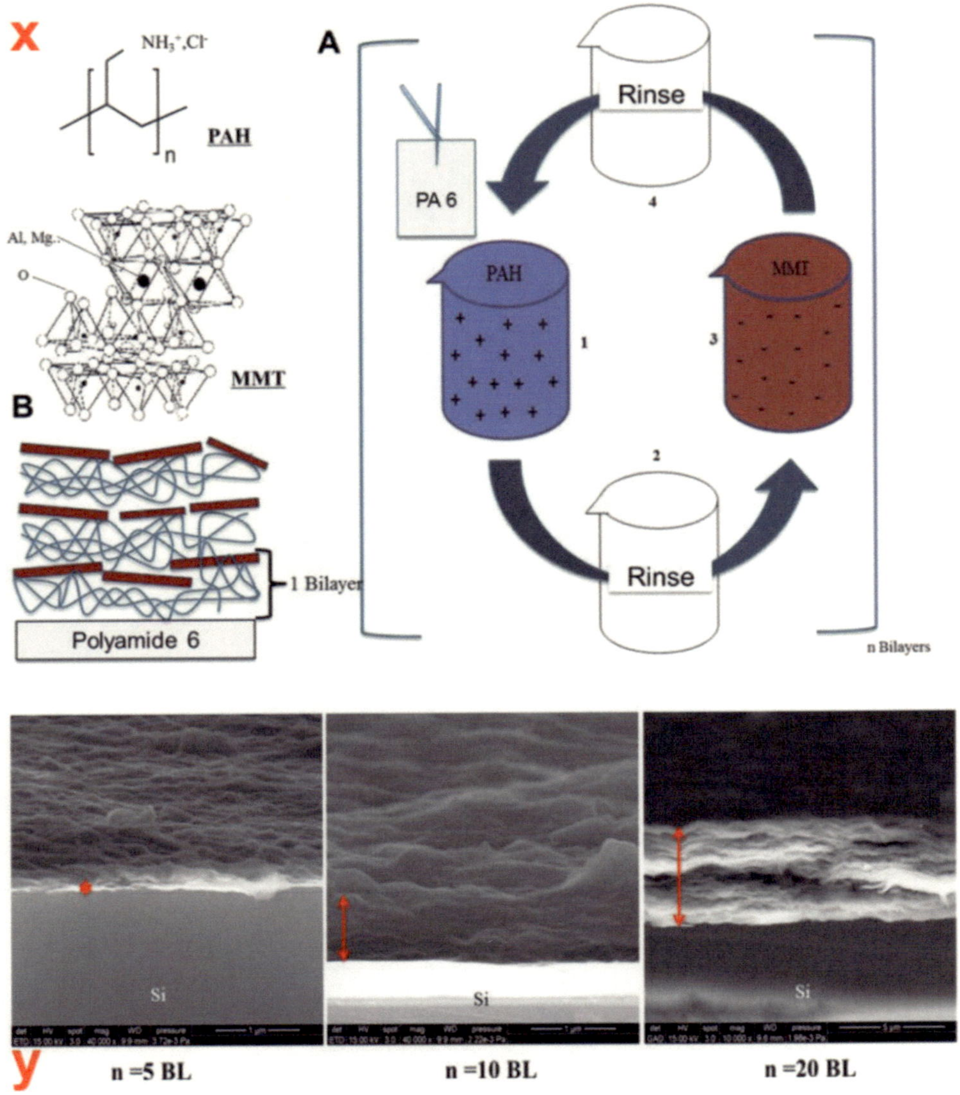

Figure 16.
x) Schematic representation of layer-by-layer deposition. We repeated the steps 1–4 until the desired number of bilayers and y) a cross-sectional assemblies (top-x) and SEM cross-section of 5, 10 and 20 BL coated samples (bottom-y) [56], copyright 2012. Reproduced with permission from Elsevier ltd.

Apart from the LbL assembly, a sol–gel technique using DOPO-VTS and TEOS is also adopted in developing flame retardant finishes onto the nylon 6 fabric surfaces [57]. The co-application of silicon compounds with a DOPO derivative ensures added charring while providing thermal stability to these residues, which acts positively to stop the melt-dripping of nylon fabrics.

Moreover, flame-retarded nylon 6 textiles with a knitted structure is developed using a different technique [58]. Here, a blending technique is adopted to mix three different components, namely nylon 6 multifilament, commercially available nylon 6 polymer chips and synthesized flame-retarded nylon 66. Initially, the flame-retarded nylon 66 is developed through the co-condensation reaction of hexamethylenediammoniumadipate (AH-salt) with the salts of hexamethylene diamine and two different organophosphorus compounds

Substrate	FR compounds	Application technique	Main results	Eco-friendliness
Nylon 6 fabric [56]	Polyallylamine and MMT	LbL assembly	62% reduction in pHRR for 20 BL deposition	Partial
Nylon 6 fabric [57]	DOPO-VTS and TEOS	Pad-dry-cure	Impart non-dripping behavior, decrease the THR value and increase the char yield%	No
Nylon 6 fabric [58]	FR-PA66 + PA6	Blending technique	LOI up to 38%, V-0 rating in horizontal burning test	No

Table 5.
Collected results of flame retardant treatment for nylon 6 fabric.

like 3 hydroxyphenylphosphinylpropanoic acid (3-HPP) and 9,10-dihydro-10-[2,3-di (hydroxycarbonylpropyl]-10-phosphaphenanthrene-10-oxide (DDP). Here, the mixing of flame retardant compounds with varied polyamides ensures better miscibility and compatibility of the composites, which help the modified nylon 6 fiber to retain its physical properties as well as overcoming the post-synthesis shortcomings related to bleeding and higher loading of flame retardant compounds. Finally, the prepared fabric with a knitted structure using these flame-retarded nylon 6 fibers come up with a very high LOI value (i.e., 36–38) and reveals improvements in other parameters. The key results of this section are presented in **Table 5**.

5. Conclusions

The flame retardant treatment of nylon textiles is getting constant development with time and becoming more sustainable in terms of uses of bio-derived flame retardant compounds and the green application processes. However, in the earlier applications, the use of petro-based flame retardant compounds has been more prominent. Additionally, it is seen that the fiber stage finishing has been more preferable over the topical finishing in the previous applications. Nonetheless, this fiber stage finishing still exists at a limited scale while some sort of modification is initiated in the current applications to convert this process more eco-friendly. However, due to the limitation of fiber stage finishing method related to the poor mechanical or physical properties in the modified fibers, the topical finishing has become very popular in today's applications. Meanwhile, the green practices in topical/surface finishing, especially in terms of flame retardant compounds and application methods are increasing day by day. These bio-derived compounds show efficacy in improving some crucial flame retardant parameters, namely reducing pHRR and enhancing the char yield % as their polyalcoholic aromatic structure favors the condensed phase mechanism. However, they are yet to show the potency in improving LOI values and thermal stability. Even in some extent, coatings developed from bio-derived polyelectrolytes via ionic interaction suffer from poor wash durability, which need to be optimized further. In line, some sort of hybridization, both in application methods (i.e., sol–gel/layer-by-layer assembly, etc.) and in flame retardant compounds (i.e., organic/inorganic/nanocompounds) have been enormously used in these days. In a typical application, flame retardant compounds belonging to organic species produce charring while the inorganic species offer shielding effect to these char residues to safeguard the underneath textiles from further burning. Such kind of hybridization has been found beneficial in improving the charring ability of nylon textiles as the nylon polymers show poor charring due to their typical aliphatic structures. Though some recent works have already shown promise in imparting flame retardant properties to the nylon textiles considering the environmental impacts and durability issues of such finishing methods, still there are challenges to imply them commercially.

Author details

Chanchal Kumar Kundu[1,2], Zhiwei Li[1*], Lei Song[2] and Yuan Hu[2]

1 National & Local Joint Engineering Research Center for Applied Technology of Hybrid Nanomaterials, Henan University, Kaifeng, P.R. China

2 State Key Laboratory of Fire Science, University of Science and Technology of China, Hefei, Anhui, P.R. China

*Address all correspondence to: zhiweili@henu.edu.cn

References

[1] N.N. Brushlinsky, S.V. Sokolov, P. Wagner, J.R. Hall, in: World Fire Statistics. Report No. 10. Centre of Fire Statistics, International Association of Fire and Rescue Service, 2006. www.ctif.org.

[2] M. Kobes, K. Groenewegen, in: Consumer Fire: European Statistics and Potential Fire Safety Measures, Austrian Federal Ministry Labor, Social Affairs and Consumer Protection, 2009.

[3] H. Frantzich, in: A Model for Performance-Based Design of Escape routes, Technical Report, Department of Fire Safety Engineering, Lund Institute of Technology, Lund University, 1994.

[4] R.G. Gann, Estimating Data for Incapacitation of People by Fire Smoke, Fire Technology 40 (2004) 201-207.

[5] J.H. Cho, V. Vasagar, K. Shanmuganathan, A.R. Jones, S. Nazarenko, C.J. Ellison, Bioinspired Catecholic Flame Retardant Nanocoating for Flexible Polyurethane Foams, Chem. Mater. 27 (2015) 6784-6790.

[6] G. Laufer, C. Kirkland, A.B. Morgan, J.C. Grunlan, Exceptionally Flame Retardant Sulfur-Based Multilayer Nanocoating for Polyurethane Prepared from Aqueous Polyelectrolyte Solutions, ACS Macro Lett. 2 (2013) 361-365.

[7] C.K. Kundu, Z. Li, *L. song*, Y. Hu, An overview of fire retardant treatments for synthetic textiles: From traditional approaches to recent applications, Eur. Polym. J 137 (2020) 109911.

[8] A. Dahiya, M. G. Kamath, R. R. Hegde, MSE 554, University of Tennessee, Nonwovens science and technology II (2004).

[9] A. R. Horrocks, B. K. Kandola, P. Davies, S. Zhang, S. Padbury, Developments in flame retardant textiles–a review, Polymer Degradation and Stability, 88 (2005) 3-12.

[10] L. Li, G. Chen, W. Liu, J. Li, S. Zhang, The anti-dripping intumescent flame retardant finishing for nylon-6, 6 fabric, Polymer Degradation and Stability, 94 (2009) 996-1000.

[11] E.D. Weil, S. Levchik, Current practice and recent commercial developments in flame retardancy of polyamides, Journal of fire sciences, 22 (2004) 251-264.

[12] J. Green, Brominated phosphate ester flame retardants for engineering plastics, in Fire Retardant Engineering Polymers and Alloys, Products and Applications—Compounding—Testing and Evaluation, papers presented at La Mansion del Rio, San Antonio, Texas, March 12-15 (1989) 69-100.

[13] G.L. Nelson, Fire and polymers—an overview, in *Fire and Polymers II* (G. L. Nelson, Ed.), ACS Symposium Series 599, American Chemical Society, Washington, DC (1995) 1-28.

[14] A. Dasari, Z.Z. Yu, G.P. Cai, Y.W. Mai, Recent developments in the fire retardancy of polymeric materials, Prog. Polym. Sci. 38 (2013) 1357-1387.

[15] G.L. Nelson, Fire toxicity, in Fire and Polymers II (G. L. Nelson, Ed.), ACS Symposium Series 599, American Chemical Society, Washington, DC (1995) 291-292.

[16] D.W. Van Krevelen, Flame resistance of chemical fibers, Journal of Applied Polymer Science, 31 (1977) 269-292.

[17] M.S. Subbulakshmi, N. Kasturiya H. P. Bajaj, A. K. Agarwal, Production of flame-retardant nylon 6 and 6.6, Journal of Macromolecular Science, 40 (2000) 85-104.

[18] Interscience (Ed.), Man Made Textile Encyclopedia, J. J. Press, New York (1959) 143-144.

[19] W.H. Sharkey, W. E. Mochel, Mechanism of the Photoöxidation of Amides, Journal of the American Chemical Society, 81 (1959) 3000-3005.

[20] Y. Shinohara, D. Ballantine, Free radicals in irradiated nylon, The Journal of Chemical Physics, 36 (1962) 3042-3049.

[21] B.K. Kandola, Flame Retardancy Design for Textile. In Fire Retardancy of Polymeric Materials, 2 ed.; Wilkie, C. A.; Morgan, A. B., Eds. CRC Press Taylor and Francis Group: Boca Raton (2010) 725-762.

[22] J. Alongi, F. Carosio, P. Kiekens, Recent advances in the design of water based-flame retardant coatings for polyester and polyester-cotton blends, Polymers, 8 (2016) 357.

[23] E.D. Weil, S. Levchik, Current practice and recent commercial developments in flame retardancy of polyamides, J. Fire Sci. 22 (2004) 251-264.

[24] S.V. Levchik, E.D. Weil, Combustion and fire retardancy of aliphatic nylons, Polym. Int. 49 (2000) 1033-1073.

[25] S.V. Levchik, E.D. Weil, A review of recent progress in phosphorus-based flame retardants, J. Fire Sci. 24 (2006) 345-364.

[26] S.V. Levchik, A.I. Balabanovich, G.F. Levchik, L. Costa, Effect of melamine and its salts on combustion and thermal decomposition of polyamide 6, Fire Mater. 21 (1997) 75-83.

[27] F. Mohaddes, L. Wang, R.A. Shanks, S.M. Fergusson, Elevation of charring level of polyamide-6,6 films viaionic introduction of phosphoric acid and boric acid esters, Green Chem. Lett. Rev. 7 (2014) 184-190.

[28] Y. Li, K. Liu, R. Xiao, Preparation and characterization of flame-retarded polyamide 66 with melamine cyanurate by in situ polymerization, Macromol. Res. 25 (2017) 779-785.

[29] Y.Y. Li, K. Liu, R. Xiao, Preparation and characterizations of flame retardant polyamide 66 fiber, 2017 Global Conference on Polymer and Composite Materials (PCM 2017) IOP Publishing IOP Conf. Series: Materials Science and Engineering 213, 2017.

[30] L. Xiao, L. Xu, Y. Yang, S. Zhang, Y. Huang, C.W. Bielawski, J. Geng, Core-Shell Structured Polyamide 66 Nanofibers with Enhanced Flame Retardancy, ACS Omega 2 (2017) 2665-2671.

[31] J. Alongi, F. Carosio, G. Malucelli, Current emerging techniques to impart flame retardancy to fabrics: an overview, Polymer Degradation and Stability, 106 (2014) 138-149.

[32] S.Y. Liang, N. M. Neisius, S. Gaan, Recent developments in flame retardant polymeric coatings, Progress in Organic Coatings, 76 (2013) 1642-1665.

[33] B. Mahltig, H. Haufe, H. Böttcher, Functionalisation of textiles by inorganic sol-gel coatings, Journal of Materials Chemistry, 15 (2005) 4385-4398.

[34] C.K. Kundu, B. Yu, C. S. R. Gangireddy, X. Mu, B. Wang, X. Wang, *L. song*, Y. Hu, UV Grafting of a DOPO-Based Phosphoramidate Monomer onto Polyamide 66 Fabrics for Flame Retardant Treatment, Industrial & Engineering Chemistry Research, 56 (2017) 1376-1384.

[35] C.K. Kundu, W. Wang, S. Zhou, X. Wang, H. Sheng, Y. Pan, *L. song*, Y. Hu, A green approach to constructing

multilayered nanocoating for flame retardant treatment of polyamide 66 fabric from chitosan and sodium alginate, Carbohydr. Polym. 166 (2017) 131-138.

[36] Z. Xia, W. Kiratitanavit, P. Facendola, S. Thota, S. Yu, J. Kumar, R. Mosurkal, R. Nagarajan, Fire resistant polyphenols based on chemical modification of bio-derived tannic acid, Polym. Degrad. Stab. 153 (2018) 227-243.

[37] C.K. Kundu, X. Wang, *L. song*, Y. Hu, Borate cross-linked layer by layer assembly of green polyelectrolytes on polyamide 66 fabrics for flame-retardant treatment, Prog. Org. Coat. 121 (2018) 173-181.

[38] C.K. Kundu, X. Wang, L. Liu, *L. song*, Y. Hu, Few layer deposition and sol-gel finishing of organic-inorganic compounds for improved flame retardant and hydrophilic properties of polyamide 66 textiles: A hybrid approach, Prog. Org. Coat. 129 (2019) 318-326.

[39] C.K. Kundu, X. Wang, Y. Hou, Y. Hu, Construction of flame retardant coating on polyamide 6.6 via UV grafting of phosphorylated chitosan and sol-gel process of organo-silane, Carbohydr. Polym. 181 (2018) 833-840.

[40] M.Z. Rahman, C.K. Kundu, H. Nabipoura, X. Wang, *L. song*, Y. Hu, Hybrid coatings for durable flame retardant and hydrophilic treatment of Polyamide 6.6 fabrics, Prog. Org. Coat. 144 (2020) 105640.

[41] C.K. Kundu, Z. Li, X. Li, Z. Zhang, Y. Hu, Graphene oxide functionalized biomolecules for improved flame retardancy of Polyamide 66 fabrics with intact physical properties, Int. J. Biol. Macromol. 156 (2020) 362-371.

[42] D. Meng, J. Guo, A. Wang, X. Gu, Z. Wang, S. Jiang, S. Zhang, The fire performance of polyamide66 fabric coated with soybean protein isolation, Prog. Org. Coat. 148 (2020) 105835.

[43] C.K. Kundu, X. Wang, *L. song*, Y. Hu, Chitosan-based flame retardant coatings for polyamide 66 textiles: One pot deposition versus layer-by-layer assembly, Int. J. Biol. Macromol. 143 (2020) 1-10.

[44] C.K. Kundu, X. Wang, M.Z. Rahman, *L. song*, Y. Hu, Application of Chitosan and DOPO derivatives in fire protection of polyamide 66 textiles: Towards a combined gas phase and condensed phase activity, Polym. Degrad. Stab. 176 (2020) 109158.

[45] C.K. Kundu, *L. song*, Y. Hu, Nanoparticles based coatings for multifunctional Polyamide 66 textiles with improved flame retardancy and hydrophilicity, J. Taiwan Inst. Chem. Eng. 112 (2020) 15-19.

[46] M. Coquelle, S. Duquesne, M. Casetta, J. Sun, S. Zhang, S. Bourbigot, Investigation of the decomposition pathway of polyamide 6/ammonium sulfamate fibers, Polym. Degrad. Stab. 106 (2014) 150-157.

[47] R. Horrocks, A. Sitpalan, C. Zhou, B.K. Kandola, Flame Retardant Polyamide Fibres: The Challenge of Minimising Flame Retardant Additive Contents with Added Nanoclays, Polymers 8 (2016) 288.

[48] H. Wu, M. Krifa, J.H. Koo, Flame retardant polyamide 6/nanoclay/intumescent nanocomposite fibers through electrospinning, Text. Res. J. 84 (2014) 1106-1118.

[49] Y. Li, Y. Lin, K. Sha, R. Xiao, Preparation and characterizations of flame retardant melamine cyanurate/polyamide 6 composite fibers via in situ polymerization, Text. Res. J. 87 (2017) 561-569.

[50] K. Liu, Y. Li, L. Tao, R. Xiao, Preparation and characterization of polyamide 6 fibre based on a phosphorus-containing flame retardant, RSC Adv. 8 (2018) 9261-9271.

[51] E. Giebel, T. Schneck, J. Unold, M.R. Buchmeiser, Syntheses of intrinsically flame-retardant polyamide 6 fibers and fabrics, J. Appl. Polym. Sci. 136 (2019) 47829.

[52] Y. Di, X. Wu, Z. Zhao, W. Wang, Experimental investigation of mechanical, thermal, and flame-retardant property of polyamide 6/phenoxyphosphazene fibers, J. Appl. Polym. Sci. 137 (2019) 48458.

[53] S. Zhu, M. Shi, M. Zhu, Effects of Electron-Beam Irradiation Crosslinking on PA6 Fibers, Fiber Polym. 14 (2013) 525-529.

[54] S. Zhu, M. Shi, M. Zhu, Thermal and anti-dripping properties of g-irradiated PA6 fiber with the presence of sensitizers, Mater. Lett. 99 (2013) 28-30.

[55] N.K. Pramanik, R.S. Haldar, Y.K. Bhardwaj, Radiation processing of Nylon 6 by e-beam for improved properties and performance, Radiat. Phys. Chem. 78 (2009) 199-205.

[56] K. Apaydin, A. Laachachi, V. Ball, M. Jimenez, S. Bourbigot, V. Toniazzo, D. Ruch, Polyallylamine-montmorillonite as super flame retardant coating assemblies by layer-by layer deposition on polyamide, Polym. Degrad. Stab. 98 (2013) 627-634.

[57] A. Sehi, B. Tomsic, I. Jerman, J. Vasiljevic, J. Medved, B. Simoncic, Synergistic inhibitory action of P- and Si-containing precursors in sol-gel coatings on the thermal degradation of polyamide 6, Polym. Degrad. Stab. 128 (2016) 242-252.

[58] G. Mourgas, E. Giebel, V. Bauch, T. Schneck, J. Unold, M.R. Buchmeiser, Synthesis of intrinsically flame-retardant copolyamides and their employment in PA6-fibers, Polym. Adv. Technol. (2019) 1-11.